THE LIFE OF FISH

Nature's Ultimate Survivors

NICHOLAS KABLE

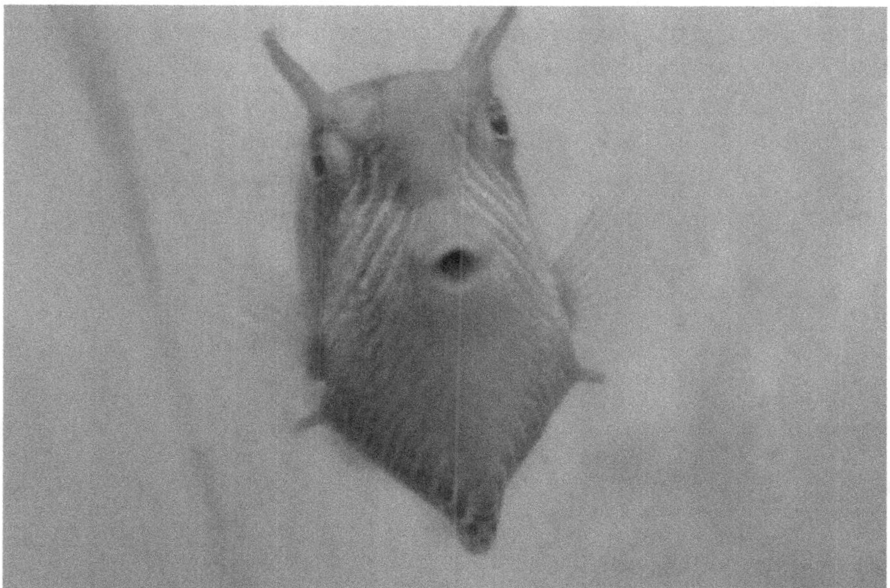

The Life of Life: Nature's Ultimate Survivors
30 August 2013
2nd version 6 March 2016
3rd version/ First published Digitial edition released July 2025

Published in Australia by Kable's Port Macquarie Printing Service
Second Digital Edition, ISBN 978-1-764663-0-5
First Paperback Edition published May 2026, ISBN 978-1-7646663-1-2
(Updated version published August 2026).

This book is Human Authored.

Contents

▼Hunting Lizardfish- ***Joe Belanger***

Introduction

Bird Wrasse- *Diegograndi*

Fish are the largest and most diverse of all the groups of backboned animals. More than thirty-seven thousand species dominate the world's oceans and rivers; more than the number of bird, mammal and reptile species combined. Some of them survive in frigid waters, cut off by from the sunlight by more than eight kilometres of ocean. At the other extreme are fish clinging to existence on mountains rising five kilometres above sea level. The success of fish in marine environments is unsurpassed by any other groups. Even the existence of such animals as penguins, dolphins and seals is only due to the presence of these marine survivors.

Despite their undoubted success the world of fish is for many of us an alien one. Besides for watching a solitary goldfish circling aimlessly in a fish bowl it is hard for us to see much of their behaviour. In part this is for a fundamental reason. Under the water we can only hold our breath for so long, in turn nearly all fish are equally powerless above water. Visibility vanishes rapidly as soon as you submerge your head; in the clearest stretch of ocean you're lucky to see 150 metres in front of you. As your ears begin to fill with water a bubble of air trapped inside them acts as a seal to avoid internal damage being caused by changes in atmosphere pressure but this same process effectively stops the spread of sound into the ear canal. So we are left deaf as well as blind. Our senses and anatomy actively work against us gaining any understanding of the marine world that covers so much of our planet and the behaviour of our most distant living backboned relatives. The unfortunate result has been a public perception of fish varying somewhere between simple cold-blooded animals with three second memories or blood driven monsters straight out of a horror movie.

Those observers who have dared to voyage beneath the waves have found behaviour as striking as that of any bird or mammal. In the last several decades the use of increasingly sophisticated equipment has revealed more about the strategies fish use to survive. Many fish have eyes sensitive to light and colours well beyond our visual capacity. Specialist cameras are revealing the individual ultraviolet patterns on reef dwellers like damselfish that previously looked identical. We are seeing for the first time the clues that fish themselves use to tell each other apart. Increasingly sophisticated audio recording are allowing us to hear what our ears alone cannot; a veritable dawn chorus being broadcast over the coral reefs, the cue that larvae of many fish use to navigate and find a place to call home. The advance in slow motion film has captured hunters that can react in a fraction of a second and found fish in the Amazon that can actually fly. In the deep ocean specialist submersibles are travelling into water with pressure several hundred times that on the surface in order to film species that have yet to be named let alone understood. These new discoveries have continued right up to the present day; in 2018 it was the first ultrasonic image of a whale shark, in 2022 it was a joint finding by a Japanese and Australian mission of the deepest fish so far known, in 2025 it was research showing the impacts that gills have on fish growth- research that completely contradicted that of only a year before, while 2026 started with the announcement that artificial light from cities was impacting the production of chemicals associated with sleep in coastal species of shark- potentially impacting their ability to compete with other shark species that are able to leave and shelter in deeper darker waters.

The wide-spread success of fish, surviving virtually everywhere that water is available, means that unlike possibly any other major animal group, new behaviour can be observed and photographed at ease not just by scientists but by anyone with the inclination and the necessary curiosity. What other type of animal is there that could be swimming around peacefully in a tank at home after being acquired via a rather boring transaction at the local pet-store that could potentially be a species completely unknown to science? For some varieties, such as bristlenose catfish or rainbowfish, that is a distinct possibility. Knowledge only grows when it is shared however, and there are a terrific amount of people- some of them scientists, others deep sea fishers, snorkelers or professional photographers- who have been generous enough to share their images, and often their knowledge, in this book. Each artist's details are listed with their work, for they are all story-tellers, each of their images contributing in their own way to bringing these striking creatures to life.

For the behaviour of many fish can be in complete contrast to general expectations. Perhaps the most striking example regards one of the most demonised of all fish, the shark. Off the coasts of Galapagos scalloped hammerheads come together in large groups each day. As the sharks jostled with each other, ramming their peculiar heads into each other's sides it was thought for many years that this was simply group courtship. But more careful observation revealed all of the sharks were females. Nor was this a hunting party, the sharks ignored any prey the pack passed through. It seemed that rather than being fearless hunters the hammerheads had gathered together to shelter from more formidable predators like the great white. But even while engaging in these safety measures the sharks do not easily cope with sharing their company. Some rather irritable individuals take their frustrations out on their neighbours which, if they are smaller, will rarely retaliate. Many years ago it was considered that fish were simply mindless robots, reacting to their instincts and senses and little less. To reconstruct an animal's behaviour in any other way was considered anthropomorphic and unprofessional. But years of study from scientists engaging, in some cases for years, with their subjects is rapidly overturning such thinking. These hammerheads display in their behaviour, personality. Some remain constantly on edge snapping at anything too close while others remain placid and if they feel threatened will slip away silently to another part of the shoal. At night when they are more alert they split up, each adapting different strategies to feed, some snapping straight at their prey others holding back feeblishly until they feel confident enough to attack.

The range of distinctive personalities in the sharks reveals a subtle difficulty in writing a book like this that tries in a few hundred pages to understand the general behaviour of a group of more than thirty thousand species. As the hammerheads clearly show for every typical behaviour there are exceptions, sometimes between different species and sometimes even between members of the same species.

The sheer number of fish has added to this confusion. Each species of animal has only one scientific name (listed in the Index) regardless of how many common names it has, which allows precise identification. But uniquely among animals and partly because of the aquarium trade many fish have been described independently and received separate (invalid) names. Even today there is vigorous dispute about the valid identify of some fish. And if scientists are not able to easily distinguish one kind of fish from another than how can fishermen, for instance, be reasonably expected to be able to identify which of their catch is abundant and which may be the last representatives of their species. Again the variety and sheer number of fish often acts against us understanding them.

This general lack of understanding has led to tragic misunderstandings. A mature female cod can produce more than nine million eggs in one spawning and the species was able to sustain vast fisheries for centuries while still capable of replacing their numbers. Overharvesting of the largest (and most fertile) individuals resulted in a dramatic plummet in their numbers, one that is yet to recover enough for commercial fishing to return to much more than a fraction of previous years. Such a trend is spreading to other heavily harvested fish such as sardines and tuna. Sharks, the whole group tarnished with the blood thirsty reputation derived from just a handful of species, have also been hunted to their very limits, perhaps even beyond them.

Thus it is becoming more important than ever to understand fish not just because of their extraordinary behaviour but to ensure the survival of those species which are most vulnerable. Far from the two-dimensional simple stereotype they are usually represented as fish are a highly variable group; some with intricate social lives, others that spend their lives as warm-blooded hunters and in certain cases individuals whose very lives rely not just on instinct but on precise timing and planning. Even the humble goldfish is proving to be much more than it appears.

Chapter 1. Tides of Change

Breaching Basking Shark, Atlantic Ocean, East of Delaware, USA - *Jeffrey H Skevington*

In spring each year the coasts of Cornwell see one of their largest visitors arrive. Silhouetted beneath the waves of murky green some unusual giants begin stirring. From a distance they look like torpedoes; that is until their dorsal fin breaks the surface. They cruise through the water with mouths wide open, most of them at least 10 metres long. The largest may reach nearly 15 metres. Some of them may even jump clear out of the water. But these are no whales. As the rows of gills behind their great mouths clearly show these are fish. Basking sharks are the largest fish in Europe and the second largest in the world.

But these sharks are not hunters but plankton feeders and they are well and truly ready to feed. The momentum of the shark forces water through its open mouth and back out its gills. The gills are filled with five thousand rakers, intricate branching structures coated with mucus. As the water passes through the gills the plankton gets lodged within the rakers, eventually to be swallowed by the basking shark's gargantuan mouth. The sharks aren't particularly fussy in how they get their meals though. They will even summersault under the water; doing so in a most lethargic manner. It is not uncommon for them to drift just below the surface, basking in the sunshine, hence their common name.

Sifting tonnes of water each day takes its toll though. Larger invertebrates, particularly crustaceans like krill, are prone to catch in the gill rakers and damage them. Even the tinier food also contributes to the wear and tear of the rakers. But the bigger concern for the shark's survival is less the inefficient gill rakers and more of the resultant impact on its ability to breath.

Gills are essential in the survival of this massive fish. But they come with a cost, one that has only recently been recognised. The problem with gills is that they are basically two dimensional structures trying to support the respiration needs of an animal which is entirely three dimensional. As a fish grows larger its gills gradually become less efficient until, after a certain size, it would be unable to breath. It is this gradual suffocation that triggers all fish to stop growing and mature, allowing them to become sexually active. In cold waters where absorbed oxygen is at a higher volumes fish like basking sharks are able to grow to a larger size than its coastal counterparts. Even so this shark has developed special strategies to support its great size.

As summer comes to a close the plankton begins to die off and the six tonne sharks, seemingly overnight, vanish. Their disappearance isn't restricted to Cornwall, other sharks are doing the same in places as distant as Japan, New Zealand, Ireland and South Africa. For many years the vanishing giants were assumed to head out to hibernate in the ocean depths and likely shed those massive gill rakers as well- after all they would hardly be able eat while they were practically asleep. It was a reasonable hypothesis; all that it required was for the sharks to be electronically tracked and for it be confirmed. Twenty five basking sharks off the coast of Cape Cod, in Massachusetts, were tagged and tracked by satellite and the scientists waited; eighteen of the tags would end up giving data.

One of the attached tags only lasted twelve days but most stayed on for 200 days, for one shark they were lucky enough for its tag to stay in place for a remarkable 423 days. The sharks weren't interested in simply proving anyone's pre-established assumptions. While migrating most did dive deep below the oceans, up to a kilometre in some cases, but they continued feeding. The gill rakers it turns out are not shed all at once but gradually so that the shark is always capable of eating when the opportunity arises. Some of these American individuals travelled to Florida, or Carolina, others to Puerto Rico. Most surprisingly, one even made it all the way to the Southern hemisphere- staying for close to a month in Brazilian waters near the Amazon River mouth, before continuing south until the tag eventually fell off.

European basking sharks show the same diversity of behaviours; the majority migrate north to south along the continental shelf but with others making massive journeys of more than 9,500km across the Atlantic to Canada. They behave differently in these warmer waters however. Back in the temperate waters off Cornwall and Cape Cod, the basking sharks will feed near the surface, they may even jump right out of the water- the largest fish to do so and the cause of Britain's only known shark fatality when one breaching shark landed on and capsized a dinghy in the 1930s. But in these warm tropical waters they are far less noticeable, spending 80% of their time in deeper cooler waters, where they can breathe easier.

Oceans, due to their size and the fact that sunlight typically only penetrates and heats the first hundred metres or so, have significant supplies of cold and oxygen rich water, particularly in their depths. Powerful currents continually bring these colder waters towards the surface allowing the seas to support fish at sizes much larger than those inland. For many years it was thought that it was simply living in a smaller habitat with a limited food supply that kept freshwater fish so much smaller. But a different interpretation of the importance in gills in limiting adult size has turned this theory on its head.

It is not the lack of food but the lack of oxygen that generally restricts the size of fish in warm and relatively stagnant fresh water. It is not by coincidence that each of the giant fish in fresh water relies on alternative respiration techniques as well as gills; sturgeons and freshwater rays have large holes called spiracles sucking in fresh water, catfish, arapaima and electric eels can actually surface and take a mouthful of fresh air. Whale sharks, the only fish larger than basking sharks, likewise make regular dives into the oxygen rich depths. Within the last handful of years this theory of the limiting effect of gill size has become increasingly controversial; in February 2024 it was announced disproved, by June 2025 a study of 33 different fish species came out in

support of it. What is less debatable is that there are very few fish that can survive without them.

Basking Shark- ***Greg Skomal, NOA***

Gills are one of the few identifying features shared by all fish. For there are fish without fins, fish that don't have a single bone in their body, even fish without jaws. Fish do not constitute a single group. Instead they are divided into four separate classes of widely different marine animals. The oldest group is the hagfish, a group of animals which don't even have a skull. Nearly as ancient are the jawless lampreys. Dominating the ocean as supreme hunters are the sharks and rays, united by their skeleton entirely composed of cartilage. The fourth and largest of the groups, the bony fish are the most numerous vertebrates on the planet. They include not only the piranhas, eels, seahorses and marlin but other more unusual species that millions of years ago left the water to give rise to the amphibians, reptiles and eventually the mammals. But to truly understand these highly modern fish we must go back to the beginning. We have to look first at marine animals so ancient that they are not even fish, nor do they even resemble them. We have to look at the first vertebrates.

Floating in the water half a billion years ago was a creature that's descendants would change the face of the earth. Like the surrounding shoals of jellyfish it had a soft outer body. It was slim, worm shaped and by wriggling rapidly it could manage to just hold its own against the current. But unlike anything else alive at the time it had a small rigid support frame within its body. This support frame was basically a thin rod, similar in composition to the cartilage that makes up our noses, but it had potential. Most of the other invertebrates either had completely flexible bodies like jellyfish and moved passively with the tide or protected their body with an external skeleton. Crustaceans, trilobites and most importantly sea scorpions all used this second method. Sea scorpions in particular were covered in immense suits of armour and were efficient predators on basically all their competitors. Even they however, would have been wary of the largest of all these ancient hunters, Anomalocaris, a giant lobster like monster which reached the size of a man. These monsters were capable of both strong independent movement and great speed. But they also had a serious limitation. The armour could not expand to accommodate a growing animal. In order to grow these invertebrates had to shed their external skeleton in order to grow a larger one. During this process the internal body was left soft and unprotected. Get too big and an unsupported invertebrate's body would simply collapse into itself while it grew a larger skeleton. Any Anomalocaris that were born unusually large simply would not have survived the process into adulthood.

The new worm like creature would have none of these problems. Although tiny having a supporting skeleton inside of its body meant that any of its larger genetic descendants would be spared the risks that its competitors and predators like Anomalocaris had to endure. This internal rod, technically called a notochord, was the basis of a spine. So important was this development that it gave its name to the group that all backboned animals (live or dead) belong to, the Chordates. Some of its members like the whale shark are six or more times longer than the long extinct Anomalocaris and, because it is difficult to get an accurate weight of a fossil, immeasurably heavier.

There are still animals alive today that closely resemble those first ancestors. Nevertheless they are not really primitive themselves, despite this name often being applied to them. Like us and every other living chordate half a billion years of generations has passed between them and that first ancestor. Their body-plan may appear simple but its very survival into the modern era shows its effectiveness. Still genetic testing shows these are the most distant of our surviving chordate cousins. They protect their delicate notochord and indeed their whole body inside a bag-like membrane wrapped around them like a tunic. For this reason the whole group of about two thousand species are called tunicates.

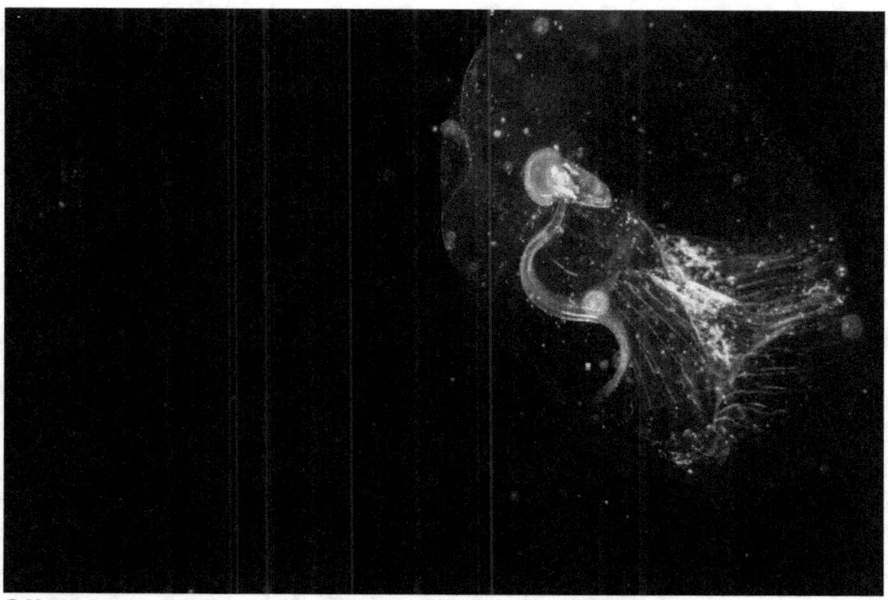

Oikopleura, Nova Scotia, Canada – *Matt Marr*

One family of oceanic tunicates, the larvaceans contain species that are basically microscopic and therefore barely known. One species however, the oikopleura, is particularly revealing about the possible appearance of the first chordates although, like the rest of its group, it is tiny. It does have two redeeming features for scientists trying to find it.

It is bright red and congregates in massive groups. Millions of these rust like specks cling together in loosely organised rows looking like a blush of algae. Under a microscope they are clearly not plants though. They look less like worms and more like futuristic tadpoles, although these are tadpoles without eyes, gills or even fins. The mouth is more like an open tube than anything else as it has no jaws to open and close it. But at such a small size oikopleura doesn't just wait for plankton to blunder into its open mouth. Instead it has a specialised organ, the rather awkwardly named, endostyle, which produces mucus in such extreme quantities that it expands into a pea sized ball, easily enough to surround the microscopic feeder's entire body. Any sizeable plankton sticks to the mucus but it also has another purpose. The continual running of mucus over the animal acts like polish on a car; reducing the risk of any debris damaging the body. The bag of mucus itself breaks down quickly and is abandoned by the tunicate after an hour or so.

Its tail is the most fish like part of the animal. The muscles are highly developed and allow a coordinated wriggle to be maintained. This wriggle drives the oikopleura forward, forcing the surrounding mucus into the tunicate's jawless mouth. As the food laden mucus is consumed, excess fluid is pushed out of little slits just behind it- gills so small that most plankton is unable to pass through. Running from its featureless head to the tail the notochord can clearly be seen. Following the notochord is a central nerve which interacts with the entire animal's body by a system of smaller nerves connected to it. Unlike all other invertebrates which carry their vulnerable nervous cord above the belly the tunicate's nervous cord actually runs above its spine. This main nerve gradually swells at the oikopleura's head into a mass concentration. Here the combined information from the oikopleura's nervous system is organised and acted on. Oikopleura has in fact one of the simplest of brains, allowing it to adapt its behaviour based on the feedback from its nerves.

There is one last, seemingly insignificant feature, to consider, the organ producing all of that mucus, the endostyle. In the oikopleura its main purpose is restricted just to that mucus but in some of its more distant relatives the organ has another critical purpose; it concentrates iodine, an essential mineral. In even more distant relatives such as ourselves a rather more complicated organ, the thyroid gland, utilises that iodine to build and release the chemicals that regulate metabolism, growth and development. This combination of gills, notochord, a nervous cord above the spine and thyroid gland is the basic pattern for all backboned animals. Even land-living mammals still have gills for one stage of their lives although this is only while developing as an embryo.

Being so small the oikopleura has an equally short life. Most only live for four to five days. In order for the genes of the tunicate to survive they must continually release sperm and eggs into the water. Surprisingly they have separate sexes unlike most invertebrates which are hermaphrodites. Having separate sexes reduces the rate of successful reproduction but this is somewhat compensated for by living in such large groups and it does have an added benefit of increased genetic diversity amongst offspring. Genetic testing has shown that the microscopic oikopleura and its close relatives are the most distant of the tunicates. The vast majority of the tunicates have abandoned the tadpole shape and have begun instead to live a lifestyle more like that of jellyfish and primitive plants. Salps are some of the most common and widespread. Most of them are a couple of centimetres but one oceanic species is about thirty centimetres, the same length as a ruler. Adults salps look less like tadpoles and more like jellyfish without the tentacles. Unlike jellyfish they propel their movement with muscles, which are layered in five bands along their body. Salps don't have tails so they can't rely on these for movement. Instead running inside their mouths are a series of specialised hairs called cilia. By contracting and relaxing their muscles the salp uses the cilia to suck water into its mouth. By ejecting this water straight out of the anus salps are able to force themselves forward. Tiny invertebrates also use cilia to move but the salp is the largest of any animal to use this method. Mucus lining the mouth and gut allows plankton to be caught during the process and feed the animal.

The larger species of salp live most of their life as solitary animals, sometimes being capable of producing young without even mating- each offspring conjoined to its parent until it is old enough to split off and live independently. Using this technique the Thalia salp has the most effective reproductive strategy of any chordate; in a single hour it can increase its length by ten percent by this process of off-shooting miniature clones. This adaptability is important because in the vast depths of the ocean it can be hard to come across food let alone another member of your own species. Come a high quantity of food however and these solitary salps produce offspring so quickly that a patch of ocean can quickly be filled with them. Individuals of smaller species of salp often join together and as different groups pass each other they form together into vast chains stretching out for several metres. By clinging so closely to each other that they look like one vast jelly like animal they dramatically increase their feeding efficiency; any plankton that misses the mouth of one animal is unlikely to avoid that of its neighbour.

Salps' bodies are contained within a bag, but unlike the mucus housing of oikopleura the bags are secreted by a hard chemical. Called tunicin, it is similar in structure to the cellulose in plants. It is the high concentration of cellulose in grass that makes it so unpalatable to most mammals. All tunicates, with the exception of oikopleura, produce the equally inedible tunicin, unlike every other creature in the animal kingdom. The bag of tunicin that salps live in contains their whole body and is permanent unlike that continually discarded by oikopleura. For most predators the combination of tunicin, and the salps small size, make eating them more effort than its worth. The bag is not the most effective protection however. Salps float along so slowly and inoffensively that an amphipod, a large woodlouse like crustacean, is able to climb into the bag. Here the amphipod spends most of its life inside it, consuming its host alive in complete safety.

In the depths of the ocean close relatives of the salps, the pyrosome family, have found a more efficient means of defence. The most common species clings together in large groups. The individuals combined form the shape of a long vase and by beating little cilia on their gills simultaneously they are able to move forward. In this shape the pyrosome colony can catch plankton easer as the water is funnelled through the increasingly narrower group until it reaches the closed end. Individually the pyrosome are tiny, much smaller even than salps, but when in formation the colony can stretch over nine metres long. During the day the enormous colony stays in the blackened depths where it won't stand out. At night it has to harness its group's combined swimming power to travel to the surface to feast on floating plankton before returning to the depths at sunrise. The colony also relies on teamwork for defence. Like the salps the bag that the individual pyrosome is contained in provides basically no protection. The pyrosome also has its own fierce predator, the leatherback turtle; a reptile known for its immunity to jellyfish poisons let alone flimsy pyrosome bodies. With its beak the leatherback can easily tear apart a colony of pyrosomes. But the leatherback has its own predators, sharks that can easily disembowel a turtle or simply rip a leg off.

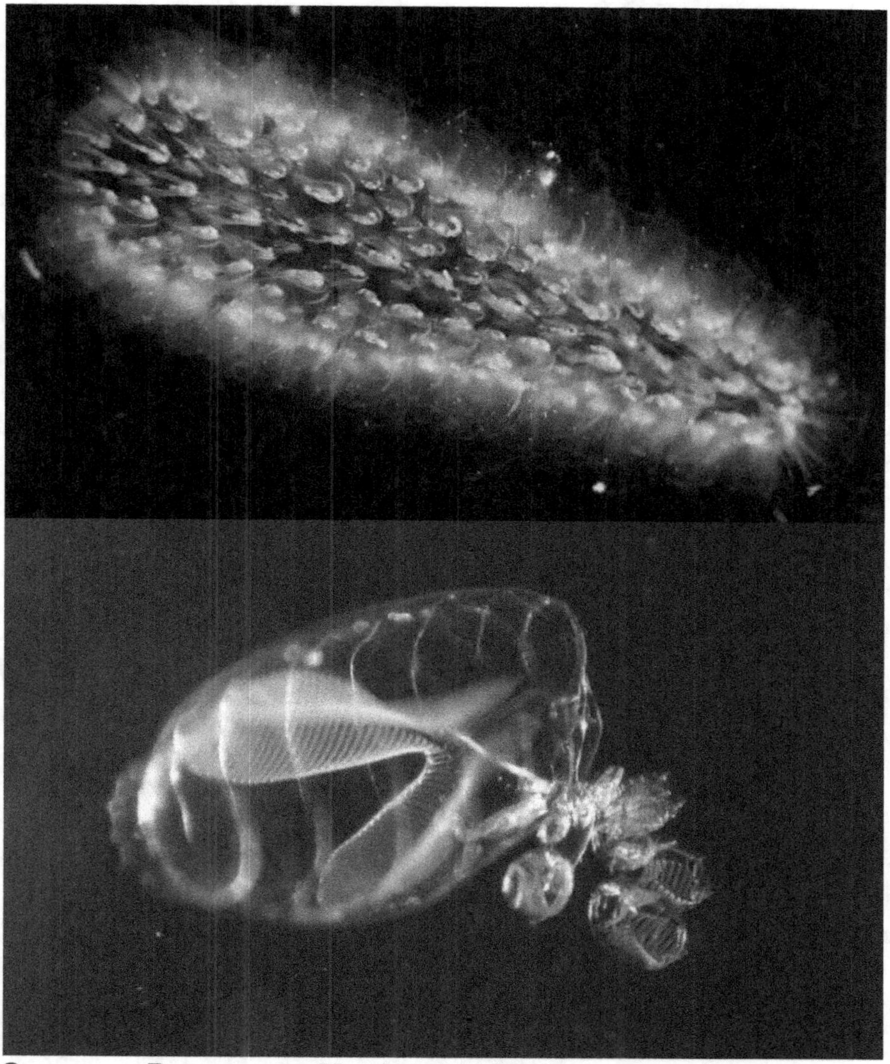

Common Pyrosome, Mabini, Philippines- *Camilla Caton*
Doliolum salp budding juveniles, Saronida, Greece- *Vasilis Stergios @vstergios78*

The leatherback relies on hunting in the darkness of night for its own security. When touched pyrosomes produce a bioluminescent green light that stands out in the darkness. Huddling so close together makes it hard for a leatherback to touch one pyrosome without alerting the whole colony of the attack. When this happens the colony produces a simultaneous flash of light that can be seen for some distance. The blind pyrosomes aren't able to see the light but the leatherback certainly can and soon retreats back into the safety of the darkness. It is not without reason that Pyrosome means 'fire body' in Greek.

Occasionally pyrosomes gather in masses below the surface during the day. Their vivid bodies can easily be seen not just from a boat but from small aircraft floating overhead. Some groups can grow large enough they can be seen by satellite. Why the pyrosome colonies choose to simultaneously reverse their pattern of sheltering by day is still unknown. Unfortunately we know more about earthworms than we do about our distant chordate cousins.

It is not the adult tunicate though that is of most interest to scientists. Despite their complicated range of behaviour no adult salp or pyrosome has a spine, brain, nervous cord or tail. But the juvenile can be a different matter. Reproduction in tunicates has best been studied in the Doliolum salp. These salps have two methods of reproduction which are alternated between generations. On reaching maturity some adult Doliolum begin to develop a tumour like growth on their underside. Soon its belly is covered in the little projections sprouting up like mushrooms. Slowly the projections slide to the base of the salp as even more of them appear on the underside. Now the projections begin to change shape and grow stalks which suck up food and feed the morphing animal. More growths appear on the disfigured salp and they continue to move up its body. Some of these growths travel up the stalks, away from the main body of the salp. Eventually the stalks reach such a size that they break away from the salp. Once separated the stalks can clearly be recognised as juvenile salps. They continue the process of budding miniature clones and releasing them into the sea with older sections of their body. But not all growths on the salp will produce clones. Some of the growths on the stalk are actually from the original salp that hitch-hiked onto the stalk before it broke away and these growths are able to reproduce sexually; releasing masses of eggs and sperm into the ocean.

The offspring that hatch from this spawning are not mushroom or stalk like. They don't even look similar to adult salps. Instead they look like oikopleura minus the protective mucus ball, in essence miniature tadpoles. They move forward with vigorous flapping of their tails controlled by their nervous system and brain. Like all other chordates the juvenile salps still have a notochord. It is only now that the relationship between tunicates and fish can be clearly seen. But it is a temporary observation only. Once they develop into adults the wriggling tail disappears and is absorbed into the body. So too is the brain and nervous system. The juvenile salps slowly develop the tunicin bag that will eventually encase them. By the time they are adults only a few bands of muscle and their gills exist as a reminder that these animals are any more complicated than the jellyfish they resemble.

Absorbing your brain and spine makes sense for animals that mainly float in the current, as this frees up a lot of energy that can be put into rapid growth. Under the right conditions salps are the most prolific breeders of all backboned animals. So successful is this strategy that salps are becoming more common in overfished oceans and the gradually warming Antarctic Ocean. One of the salps relatives is even more energy efficient- the adult is completely stationary, only as a juvenile is it capable of swimming freely and colonising new areas.

This is the solution used by the sea squirts, the most numerous and varied of all the tunicates. Their larvae maintain the typical tadpole body shape but adults go through a different process of development then other tunicates. As the larva grows it seeks, via a process that is still not completely understood, an area with ideal combination of light intensity and steady current flow. When it finds the right spot it lands, tail first, and begins the process of breaking down its brain and spinal cord until all that is left is a bag like creature, perched atop the seabed. The adult loses control of so much of its body that it loses the ability to swim and remains permanently enslaved to the same spot that its juvenile brain discovered for it. One of the consequences of being tightly rooted to the ground is that waste can't be excreted though the rectum any more. The common sea squirt has had to develop a separate tube below its mouth, connected to its anus to do the job for it. It may appear to be a solitary life but sea squirts can be just as gregarious as some of their relatives. Common across the rocky shores of Britain and the Mediterranean is the light-bulb sea squirt. With its beaming body it stands out easily to any passing diver. At first sight it does not appear any more social than a number of other species in the local area. But hidden below the substrate that surrounds it is a little outgrowth that runs like the runner of a plant from one to another. Rather than being isolated, with these connections, the tunicate can share crucial nutrients and other resources between each other.

In colonial sea squirts the relationship has become even more intricate. In this case each animal has its own mouth but a single excretion tube is shared by the group. One of the most common of these species clusters together so closely that it looks like brain coral, with each of the animals positioning its body as close to the communal vent as possible. Where conditions are right these sea squirts form groups so large and dense that other competitors are crushed out of existence. Other colonial sea squirts have formed additional alliances. In the sunlit waters the opportunities for photosynthesis, the process of extracting energy from the sun, is ideal. This is something that sea squirts being animals are not able to do on their own. The coral which they resemble so closely are also animals but by sharing their body structure with algae are able to take advantage of this resource. Sea squirts with their soft tunic body also provide the ideal accommodation for algae. As the algae take in sunlight some of the excess energy produced is shared with its tunicate host.

Bluestriped light-bulb sea squirt, Murcia, Spain- *Kaila Wheatly Kornblum*

In becoming a virtual greenhouse the sea squirt is the only backboned animal that has formed such an intimate relationship with algae. Living under such conditions some colonial sea squirts have grown into animals so different from their active vertebrate relatives that they could understandably be misidentified as being plants rather than animals. One side effect of housing algae is that it causes the sea squirts to take on some of the colour caused by all plants as a process of photosynthesis, green. With a constant supply of energy available filtering the water around the sea squirt has changed from valuable feeding to little more than a means of respiration. It is indeed only the gills, located around the mouth in the adult sea squirt, that remind us that this creature is actually more closely related to us than any dandelion or cabbage. This is what can result of a vertebrate left with an endless food supply, no need for a brain or spine: an animal that is little more than a self-replicating marine vacuum cleaner.

Not all conditions are so amiable for sea squirts. Closer to shore, with unpredictable violent waves smashing against the coast solitary species like the common sea squirt reappear, clinging on for their very existence. After the tide drops existence become even more difficult as the sun begins to evaporate any moisture left on the sea squirts. A salty mist blowing in from the ocean stops them from drying out completely but even this is infrequent. The common sea squirt however holds its own water reservoir. Unlike its colonial relatives this solitary species still has some muscle control and as the water leaves its shelter each day the animal begins to close up its head and excretion tube. In the process it keeps a ready supply of liquid trapped inside it, enabling it to keep its internal body moist and functioning until the tide returns. Only then can it resume filter feeding.

There is one sea squirt that can't afford to be this peaceful. In the depths of the ocean no light penetrates. Without this essential source of energy the staple diet of most solitary sea squirts, plankton, is also sparse. Animals here have to be able to survive on whatever they can snatch. The sea squirt that lives here, aptly called the big mouth tunicate, has become a predator. Its head fans out over its body so that it looks almost like a mushroom. Unlike other sea squirts the Big Mouth Tunicate has complete control of opening and closing its mouth. This is essential for its primary prey like shrimp have quick reflexes and won't be passively swallowed like plankton. When a shrimp approaches the Big

Mouth gapes wide, allowing the change in water suction to bring the meal inside its throat. Once caught it can snap its mouth shut fast to prevent its prey from escaping. Speed is essential for this animal's survival and the spoils have allowed it to grow profusely, at nearly thirteen centimetres long it is by far the largest sea squirt.

The Big Mouth has another advantage over its sea squirt cousins, it can move. By leaning forward it can stretch out and snatch up creatures that otherwise would be well beyond its reach. But the Big Mouth is still firmly attached to the ground and can only lean so far. Like all sea squirts the only time when a Big Mouth has complete independence over its movement as a juvenile.

The tunicates have distant relatives that are rarely stationary instead, like Oikopleura, the adults are free swimming animals with notochords. Indeed, their appearance leads them to be considered, mistakenly as it turns out, to be the closest relatives of backboned animals. Surprisingly genetic evidence has shown that the tunicates and vertebrates are each other's closest relative; but this other group of species illustrates better than any other the changes that most likely occurred in the development of the vertebrates. These are the lancelets. But, unlike their tunicates cousins that split off to become brainless spineless filter feeders the ancestors of the lancelet remained active, at least for some of the time, as adults.

At first glance a swimming lancelet could easily be mistaken for a fish. It moves its body in the characteristic s-bend movement that most fish have when flexing their spine. But the lancelet does not have a proper spine, just a soft notochord like Oikopleura. It also stays resting for the most part on the sand, if not entirely buried in it, and one good look at its head allows it to become immediately clear that this is no fish. It lacks any eyes, nostrils, ears or even jaws. Now only its gills give it some resemblance to a fish. Unlike sea squirts the brain is also maintained into adult life by lancelets, although that is certainly not visible to the naked eye.

Not having jaws means that the lancelet cannot prey on large shrimps like the big mouth tunicate. But it does have something else, cilia. Its relatives may use these hairs for movement but the lancelet uses these for an entirely different purpose. By waving these cilia together the creature pumps water down its throat and with it any suspended particles.

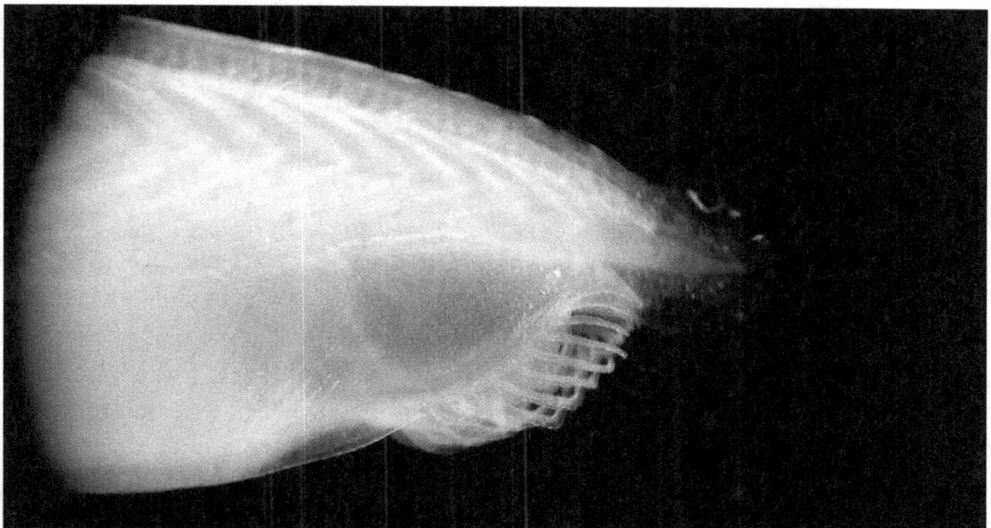

▲►European Lancelet close-up of mouth and swimming, Jersey- *Chris Isaacs*
New Zealand Lancelets, Wellington, New Zealand- *Luca Davenport-Thomas*

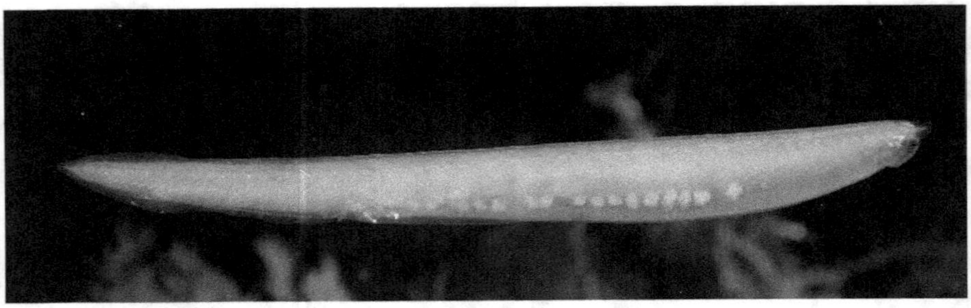

A bag like membrane inside the lancelet's mouth catches these edible particles while the rest of the water is either expelled through the gills or is forced out its vent. The lancelet is small enough that the oxygen being absorbed through the skin is enough to meet its minimal energy needs; leaving the gills free to prioritise feeding and not breathing- meaning it doesn't have to pause while feeding to catch a breath. This unusual creature doesn't even have a heart. The lancelet's blood simply flows throughout the body. It spends most of the time with just its head poking out the sand, sucking in water. The lancelet does have one, very small, warning system. At the top of its head are two sensitive areas that react to any changes in light, like the shadow cast by a passing fish or octopus. These unusually delicate pieces of skin give the lancelet a small visual understanding of the world around it and are called in honour of this, eye spots.

The surprisingly enterprising creature also makes its own colour with fluorescent green patches on its tentacles and another two below the eye spots. It is unclear what exactly the patches do but there is a possibility it attracts plankton towards its mouth. Meanwhile the lancelet is looking, as best as it can, for any changes in shadows or passing movement that could indicate its own predators. Its dull colouration blends in with the substrate but sometimes that isn't enough and it has to flee. When it does so it swims smoothly away with an ease you would not expect from an animal that is so visually primitive. At this time it is easy to imagine you are watching the movement of the first ancestral fish rather than a clumsy evolutionary dead end hovering millimetres above the mud. Despite its peculiarities the lancelet is one of the closest surviving relatives to the vertebrates and in both its anatomy and behaviour we might be able to see the blueprint for all the backboned animals alive today. Indeed the fossils of the first fish ever found, about five hundred million years ago, bear a striking resemblance to the living lancelet, with their slender spines and simple mouths.

The processes of time has acted against the likelihood of any solid fossil evidence of the first fish surviving to the present day. For any fossil to form, conditions have to be ideal, something that becomes increasingly unlikely the further back in time you go. Firstly the animal's body has to avoid being eaten or having its body broken down by bacteria or scavengers. Next it has to be covered in enough sediment to preserve it but not enough to crush it out of existance. Here, surprisingly quickly, minerals from the surrounding debris such as limestone begin to replace the chemical compounds of parts of the skeleton. This can happen in as little as a few decades. But the rock encasing the fossil takes millions of years to harden enough to preserve the animal properly. Fossil hunting is generally restricted to land meaning that for remains originating from the bottom of the ocean to be recovered these same fossils must have withstood a change of time during which the oceans themselves have moved and the sediments encasing them have forced their way up to the surface of the earth.

There is one further reason why a creature as ancient as the first fish may not appear in the fossil record. The perfect place for fossilization is where the water has been so polluted and existance harsh enough that bacteria is unable to occur in high enough numbers to damage fresh carcasses. Such environments are unlikely to attract fish in the first place which by their very nature need an environment with oxygen rich water to survive. Yet occasionally an environmental disaster like a volcano erupting and covering a section of ocean with ash and pumice may be able to preserve such ancient histories.

Such an event seems to have happened in a time period called the Cambrian, about 530 million years ago in Yunnan Province, China. Representatives of about 180 different species were killed and preserved by the event, most of them clearly invertebrates but some were obviously something else altogether. Genetic analysis indicates that fish had already split from primitive chordates like sea squirts and lancets by 550 million years ago, so more than enough time had passed for the two groups to begin to look different enough to be distinguished.

More than five hundred different individuals give us a very clear insight into the body plan of what was long considered the earliest fish, Haikouichthys. Recently scientists have reconsidered the species classification and it is now instead considered to be a primitive chordate rather than a fish but it still provides a clue as to what the earliest fish possibly looked like. It was an active creature with lines of muscle blocks running down its body to precisely control its movement. It was one of the earliest backboned animals with fins, with one located above the spine, another under the abdomen and perhaps a third on its tail (fossils are yet to be found to conclusively determine whether a tail fin occurred or not.) The rows of gills on Haikouichthys were prevalent and likely used more for respiration than for feeding. To really appreciate what makes this early creature so special though you have to look at its spinal cord. In some specimens parts of the notochord are molded in a similar form to the vertebrae of modern jawless fish, rather than the simple cord shape seen in tunicates and lancelets. The skull is also solidified providing extra protection for the brain and allowing it to focus on sensory information, potentially using vision, scent and sound to base its decisions of survivals off. Haikouichthys was, put simply, one of the most advanced vertebrates of its time, or at least one of the most advanced we have yet uncovered.

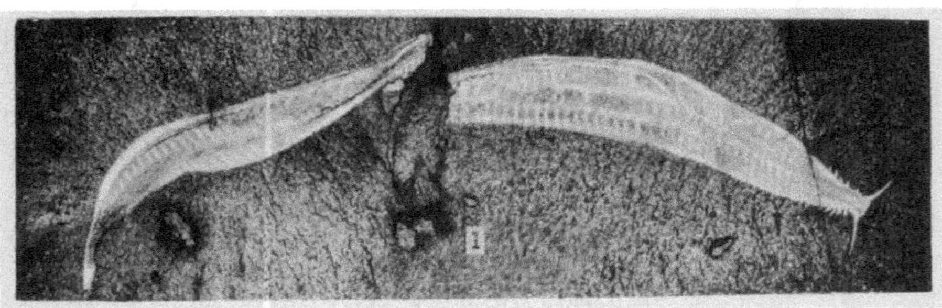

Pikaia- ***Insert from Walcott Cambrian Geology and Paleontology II***

At the time all of the continents were joined together and China's closest neighbours would have been North America and Greenland. Fortunately whatever conditions preserved so much of the landscape in Yunnan Province seems to have acted here as well. The fossil beds of Greenland have not been examined as closely as its Chinese contemporie but in North America things are a little different. The most complete Cambrian fossil beds here are about fifteen million years younger than those in Yunnan Province, concentrated in the famous Burgess Shale formation. It may be younger but many of the species here are basically identical to those in China, such as

Anomalocaris and many of the trilobites. Haikouichthys has not been uncovered here yet but, due to the closeness of the two continents, this is more likely to be due to the small size of the fish rather than a complete absense. Instead encased inside the rock is what looks remarkably similiar to a lancelet, the finger sized Pikaia. It looks stunningly like its modern doppelgangar with its undefined head, gills, lack of fins and notochord. The most surprising difference is that in its most recent reconstruction, its backbone is now considered to have been on the underside of the swimmer's body, with its digestive track located where the spine is in every other backboned animal.

It used to believed that fish and all the other vertebrates descended from an ancient lancelet. The presence of fish in fossils deposited several millions of years before the first known lancelet makes that highly unlikely. Now more sophisticated techniques comparing the genetic difference between groups of chordates is shedding more light on the evolution of the fish. Current evidence indicates that 550 million years ago an animal similar to oikopleura founded three major groups of descendants. The first group to separate, and the most distant, genetically, from the other two, gave rise to the lancelets half buried in the mud.

The remaining groups split from each other considerably later; the second group became the tunicates, regressing to a filterfeeding lifestyle- either stationary like the sea squirt or actively swimming through the plankton like the salps. It is the third group that will occupy the rest of the book for these chordates gave rise to all of the backboned animals alive today, not just in the sea but on the land. So it is not the case that fish evolved from some ancient lancelet, they evolved their body shapes separately, not as the result of one being descended from the other. This does not deny the fact that creatures like the seasquirts and oikopleura nonetheless shed light on our joint ancestors.

Some fish are as bizarre as the tunicates and lancelets that so closely resemble their ancestors. Indeed, they are so other-worldly that is debatable if they should even be classified as backboned animals at all or instead as invertebrates along with the insects, starfish and tunicates.

The Pacific hagfish is one of these oddball fish. Unlike Haikouichthys they don't have eyes; instead they have eye spots, like lancelets. Unlike any other fish hagfish don't even have a spine or fins. Instead they have a soft notochord to power their movement through the water. The brain however is protected inside a solid skull, like every other fish and all land vertebrates. The skull has a single vulnerable area, an exposed pore below the eye spots that allow volatile chemicals in the water to seep their way into the brain. With this nostril hagfish can process the scents that flow past them, sniffing out polychate worms, their main meal, and plunging into the sand to drag them out. Remarkably the hagfish can still breathe even as the clogging mud is sucked into its mouth with its prey.

Running across the front part of the hagfish's body is a series of gills. They are not the delicate fringes of skin in other fish. Instead they open up like little pin pricks across the front half of its body. The gills are anatomically separate from the mouth meaning that the hagfish is not able to gulp in additional water at leisure. Consequently the hagfish is not able to extract enough oxygen from its surroundings to engage in high energy action like pursuing prey, leaving anything faster than a polychate worm off the menu. But having a respiration system separate to the mouth means that when times are right the hagfish can continue feeding non stop without needing to pause for a breath. This is not nessecary for smaller meals but occasionally hagfish stumble across larger delicacies. Again it is their single nostril that leads them to their meal.

There is one meal in particular that seeps an odour through the ocean, one that can keep the hagfish feasting for months. A whale carcass. The hagfish's noodle-like body is simply not designed for speed; in a reef competition for meat would soon result in starvation. But in the fozen depths its lifestyle has advantages, it uses little energy. While other fish would starve to death in the darkness the hagfish can simply wait for a meal, sometimes for a couple of months at a time. When the hagfish reaches the carcass their main competitors will be each other.

Whales have one major problem as a food source. Their solid bodies are covered by equally tough skin. Encrusted barnacles add extra defence to the carcass. The hagfish has another problem. Although it has a solid skull, like the lancelet, it has no jaws. Instead it has a wide opening at its skull lined with teeth, to tackle the corpse with. It doesn't even have proper teeth instead what it has in its place are relatively soft and pliable spinks made out of keratin, the same material that hair and finger nails is made of. These fake teeth are positioned on what in other animals would be called a tongue. By being brought together they do cut into their prey but this is effective in only a crude way. But having no

spine is inclined to leave you with something else- extreme flexibility and the hagfish know how to use this to full advantage.

After gripping onto the whale the hagfish starts to wriggle around tightly. It needs leverage but the only anchorage available is itself. Wrapping its tail around its body it starts to tie itself into a knot. Gradually it wriggles the knot down its body until it can brace itself and pull back ripping a piece of skin away with it. Having no jaws means it is unable to chew so it has to suck on whatever piece of meat it grabs until it is small enough to swallow whole. The hagfish also has to untangle itself out of a knot. For this however it has a special strategy. Its body is lubricated with slime glands and with continued pressure it is able to wriggle and slide itself free.

Hagfish have another even more, if possible, unusual feeding mechanism. All backboned animals inevitably have an exchange of liquid and nutrients between their body and the outside world. Even people can notice this effect as their hands and feet begins to wrinkle up after being immersed in a bath for an hour or so. Most fish have their bodies covered in dense scales in order to reduce this effect but they also regulate the distribution of nutrients in their urine and other discharges in order to keep the working conditions on their body as regular as possible. Hagfish take none of these precautions and hence put their bodies through a range of extreme conditions that basically no other fish, let alone any other backboned beast, could cope with. But one advantage of this is that the hagfish can rest on a carcass and continue to absorb its meal through its skin, saving its energy in the most efficient of ways.

Whales being so large are prone to attract a lot of predators, and the corpse is soon swarming not just with hagfish, but conger eels, sharks and amphipods. Even green moray eels from the coast may venture into the depths for the feast. Not all are content with rotting whale flesh and for some the temptation of a fresh hagfish is too much. It is a temptation however that could prove fatal.

Adult hagfish are too large for most predators to swallow whole. Instead a potential predator, like a shark, will normally try to bite it into smaller sections before attempting to swallow it. As soon as its skin is touched however the hagfish starts to produce copious amounts of slime. Before the predator's teeth have even abraded the surface the skin is frothing. It is now that the hagfish's spineless body can truly come into its own.

◄Sixgill Hagfish, Steenbras Deep, Cape Town, South Africa- ***Peter Southwood***
◄Pacific Hagfish - ***Linda Snook, Monterey Bay National Marine Sanctuary***

As the trapped scavenger starts to squirm around it simultaneously slides the slime up its body towards the shark's open mouth. When fished out of water the hagfish produces enough slime to fill a twenty-litre bucket in minutes, underwater the fish's technique is equally effective. Wriggling around forces more slime out of its pores, making the fish increasingly difficult to grip onto. The slime has another quality. Exposed outside the hagfish's body it starts to stabilise, in the same way as a spider's silk. If the shark doesn't let go quickly enough its mouth is soon clogged together. As the shark struggles the still-soft slime is forced back across its body and it starts to harden, forming a seal over the would-be-killer's gills. If it is unable to free itself quickly enough the shark is at risk of asphyxiation and the hagfish is left with another potential meal.

The hagfish appears to our eyes as an obscure oddball. In the darkness it is easy to remain ignorant about the number of hagfish and hence their significance. Yet there are more than fifty species of hagfish known despite the relative inaccessibility of its environment. This is a larger number of species than there are of deer and few people would call those land animals insignificant. Originally all hagfish were lumped together in a single family now comparative genetic and anatomical study, particularly of different gill formation between groups has shown there are at least two different families if not more. It would be not be surprising if these different families likewise have markedly different behaviour between them, including behaviour that we are yet to discover. The hagfish may appear weird but it is also highly successful. But even with such a large number of species one part of its life history is still virtually unknown, its reproductive life.

Unlike man y other primitive chordates the hagfish has separate sexes. Being hagfish even this isn't entirely clear-cut and they seem to be able to change gender during their life, but can only reproduce as one sex or the other at any time. Although how they distinguish the sexes in the darkness is largely unknown it is likely that their advanced sense of smell guides them in some way. What happens with the rest of their life story, if the hagfish have any courtship ritual or other peculiar reproductive behaviour is unknown; until it comes to their eggs which can be found lying on the seabed. They are large and yellow, like oversized peanuts, held in place by small tendrils that grip the ground. Growing inside the eggs are identical replicas to their seniors. When the hagfish break out of their shelter they are basically miniature adults. They can also begin hunting and seeking out food the same as their parents. But there is an advantage to juveniles being different to their parents. Like the young of the sea squirt it allows the animal to take advantage of different opportunities throughout its life. It also allows it to avoid competition with more experienced adults of its species over similar food. The hagfish has a relative that has taken this evolutionary route.

Wriggling in the mud of a European lake is an ammocete. Unlike any chordate we have encountered before this lively creature does not survive on straining plankton, scavenging or even hunting down worms. It is primarily a vegetarian feeding blindly on rotting vegetation and algae growing on the rocks. To break down this food, requires a strong digestive tract filled with bacteria and corrosive chemicals, for the cellulose that binds plants together can be hard for large reptiles and mammals to break down let alone small fish. But even an acid filled intestine (for the ammocete has no stomach) requires oxygen to function. So deep inside its body is a powerful muscle that pumps oxygen filled blood around its body, using specialised blood cells that turn red as they fill with oxygen. As they make their way around the body the red blood cells begin to lose their oxygen and some of their colour. They also become at risk to infection and bacteria. To protect these cells are smaller white blood cells that gobble up any abnormalities in the blood stream. The oxygen carrying blood returns to the muscle via a second chamber to become filled with oxygen again and continue around the body. The ammocete is running its entire body on this tiny beating heart, a muscle that has become not just a convenient aid in respiration but the creature's most critical component, a pattern which it shares with virtually all of its more advanced relatives. Hagfish also have a heart but because their anatomy already allows an easy exchange of water between the body and the outside world its internal structure is not as advanced as that of an ammocete.

That is not the only advanced feature of ammocetes; they have not only a solid skull like hagfish but an entire skeleton made out of cartilage. These are true vertebrates but they are missing something else crucial to every other backboned animal on the planet. Jaws. The ammocete has a simple straw like mouth, not much more advanced than that of a lancelet, but it can move its extended upper lip, just. It also has a set of tentacles inside its mouth that when used with its lips help to loosen algae enough to strain it into its mouth. Once inhaled the tongue secretes a specialised mucus, saliva, that helps to drag food into its gut and commence digestion. But ammocetes do not stay like this their entire life. Like the free-swimming larvae of the sea squirt the ammocete awaits a transformation which will change them into another animal entirely. After three years most species of ammocetes descend back into the ocean to complete their transition into adulthood. When they return to the streams and rivers to breed several years later they are no longer called ammocetes but lampreys.

This time the lamprey can see where they are going. Rather than using its eye spots to pick up shadows the adult lamprey has true eyes to help the fish negotiate its way around its surrounding. It may also use its single nostril to help it recognise the sweet fresh waters of its birth. In Europe one of the most common species to undertake this journey is the river lamprey. Once it arrives back in fresh water it will no longer be able to feed so in compensation it hunts heavily before it leaves the safety of the sea. Unlike the ammocete the adult river lamprey is not restricted to a diet of algae and microscopic invertebrates. The tentacles that rimmed its mouth as a juvenile harden into multiple rows of sharp teeth that spiral around the underside of the fish's head. The outside teeth are tiny, but each row inside are increasingly larger so that the inner row of teeth project like sabres. Such a mouth means that mature lampreys are not restricted to scavenging on the leftover's of more efficient predators and are able to hunt down live prey in their own right.

The mature lamprey still doesn't have any jaws but it does have a sucker disk enclosing its mouth. With this disk the river lamprey can stick against the soft body of a squid or the antennae of a crab and keep rubbing its teeth over the same spot until it can rip it apart. The sucker means that, try as it will, without the aid of limbs few of the animals attacked by a lamprey will be able to detach it. As the prey struggles the lamprey continues to pull itself closer into the flesh of its captive, gnawing continually as it slices through fat and sinew. But some denizens of the deep are too large for the river lamprey, most adult fish for instance are simply too powerful for it to subdue. Their strength allows them to outmanoeuvre and outswim the lamprey before it has a chance to attach.

Having a sucker disk may be vital in feeding but it also has a serious detractor for a swimming animal, it produces significant drag. So lampreys leave their sucker closed for the most part, only opening them just before taking a bite. The fish have other adaptations for travel. They have larger gills than the hagfish, that vaguely resemble the port holes on a ship. Unlike its relative and the lancelet the adult lamprey is unable to breathe through its skin and it relies entirely on its gills to meets its oxygen needs. Due to its increased importance these gills act more efficiently than in its distant cousin and so are reduced in number, allowing room in the body for other features such as fins. The lamprey really only has one significant fin, a dorsal fin that rises from its back, but even this single fin aids the fish in slicing through the water. River lampreys are not the only lamprey participating in such migrations.

In South America, Australia and New Zealand the pouched lampreys are also migrating inland from their home in the sea, on what can be a sixteen month one-way journey. The males produce large sacks below their mouth, giving the species their common name. It is still not known what exactly these pouches are for. It is thought that they may be used to provide protection during fighting or perhaps as part of a ritual display; another thought is the fathers use the pouches to pick up the eggs or possibly even to aerate them. Regardless all of the pouched lamprey will die soon after spawning; both male and female. Their young will be left to fend for themselves and will return to the sea once they are old enough to tackle larger prey.

In Europe there is a species of lamprey that remains in the rivers and brooks year round. The brook lamprey does not develop a sack but instead uses its mouth for its own intriguing behaviour. They begin to climb upstream a few miles before they breed, so that their offspring will eventually be washed downstream to grow in the same waters that their parents did. The males arrive first and begin to clasp large stones in their mouths, moving them to create a burrow among the pebbles. He continues in this way until a female appears and then the purpose of this incessant digging becomes clear; he has created a nest. Lampreys are not hermaphrodites; they are either born male or female and once a pair has established itself they proceed to take advantage of the male's curious construction and spawn.

Pouched Lamprey, Great Ocean Road, Victoria, Australia- *Cath Rampton*

Lampreys and hagfish may be the only surviving members of their group but they are by no means typical of jawless fish, any more than a panda could be considered a typical bear. At their peak jawless fish were some of the top predators to be found anywhere in the world. They also had to be well defended from each other. Many of them developed great shells of armour that encased their entire body. Like the lamprey eyes were an essential part of finding food and shelter for these fish but they also developed other strategies. One fish, Cephalapsis, developed sensors above the rim of the mouth, with an unusual ability. In order to control movement in a body small amounts of energy, electricity, are generated. Normally the amount of electricity involved in our bodies is so small that humans barely notice it until we touch a static object like a parked car and get a zap. In the water however electrical particles are much easier distorted and spread out. It is these particles that Cephalapsis' facial sensors appeared to be picking up. Such sensors exist today in many modern fish, most notably sharks and sturgeons.

Electrical sensitivity was not the only advancement made by these creatures.

To navigate its way around without stumbling into predators or other hazards required fish being able to respond to other sensations surrounding them. Touch is a sense shared by all surviving fish and land vertebrates but it has a substantial side effect, pain. Pain may be uncomfortable but it is possibly the most important part of touch due to its role in alerting the body to impairments and reducing the risk of an injury being exacerbated through carelessness. Because pain has such an emotive response on the body for many years it was considered to be only a higher functioning activity and required substantial scientific testing, and discomfort on the part of several participating trout, for pain to be proved unquestionably as exiting in bony fish. Touch and pain almost certainly occurs in living jawless fish and if that is the case it is more than likely that such sensation also occurred in their ancient relatives. Cephalapsis and other members of its family also found a way similar to the massive invertebrates to protect their vulnerable bodies, armour. Their bodies were encased in a solid shell of thickened scales but, unlike Anomalocaris, as the fish grew it didn't have to worry about shedding its outside protection as its skeleton still remained internal.

Another class of ancient fish developed even more efficient body armour, the Ostracoderms. But the key improvement in the group's body plan was the fins. Rather than just single fins on the tail, abdomen or back for streamlining most of these species added another innovation to improve swimming. They had a pair of fins situated on the chest (the pectoral fins) and further back an additional pair growing from the pelvis. These fins provided the fish with a higher level of control and steering ability, allowing it to navigate its way around spiny coral and poisonous anemones without being stung. In order to protect these precious innovations the fins were encased in armour nearly as solid as the rest of the body, a feature which would have slowed them down considerably. Coordinating additional fins and more active movement required something else, a larger brain. But large brains have a problem; they are incredibly hungry for energy. A diet sucked up particle by particle as Cephalapsis and many of the Ostracoderms did is far from effective at providing this.

Three hundred and seventy million years ago another incredible fossil has been found, with anatomical improvements that resolved the dietary problems faced by many of its predecessors. Dunkleosteus. Its gills had changed in shape from port holes into long, slits. This enabled them to improve their breathing and support an increasingly active lifestyle. It also allowed Dunkleosteus to reduce the number of working gills and modify one of them into a different purpose entirely, a flexible proportion of the head. Its skull was hinged on a transformed gill slit so that the lower part could be dropped and raised. By this modification the ancient fish were able to develop some of the earliest and deadliest jaws. It did not have cheeks though, so any food it caught had to be swallowed whole. Dunkleosteus did have one advantage against its prey though. Inside its mouth were solid blades capable of slicing through the armoury of any trilobites or small fish in its way. These blades were in fact almost identical to the armour that encased them. As its diet developed so did its size and this monster fish grew as large as a great white shark. Such jawed fish proved great competition for the first jawless fish. Before the newcomers reached dominance there were eight different major divisions, or classes, among jawless fish. To put it into perspective all living mammals, including marsupials and the egg-laying echidnas and platypus, all belong to the same class. All birds from eagles to flamingos also belong to a single class. Each of the eight classes of jawless fish also held species showing a great variety of body forms and lifestyles. But today only two separate classes of jawless fish still survive- the obscure lampreys and hagfish. Undoubtedly one of the reasons they continue to exist is that they tend to favour environments where there is little competition.

Dunkleosteus was undoubtedly more advanced than the initial jawless fish it preceded. But being encased in solid armour as well as providing protection also reduces manoeuvrability and speed. And solid blades are not the easiest way to eat or catch a meal. Not only this but these blades also reduced its ability to catch smaller prey, like krill. With the advent of increasingly agile prey this mighty fish was doomed for extinction. But other early fish took this change as an opportunity to thrive. There is one fish alive today that gives us an indication of what these other early fish might have been like. The bramble shark's rounded blob of a body does look similar to pictures of Dunkleosteus. It is its skin though that is dramatically different. As you approach it the skin appears smooth and unarmoured. A slide of a gloved hand against the shark will prick your skin. With bare skin a rough touch against the shark's scales could slice your hand open.

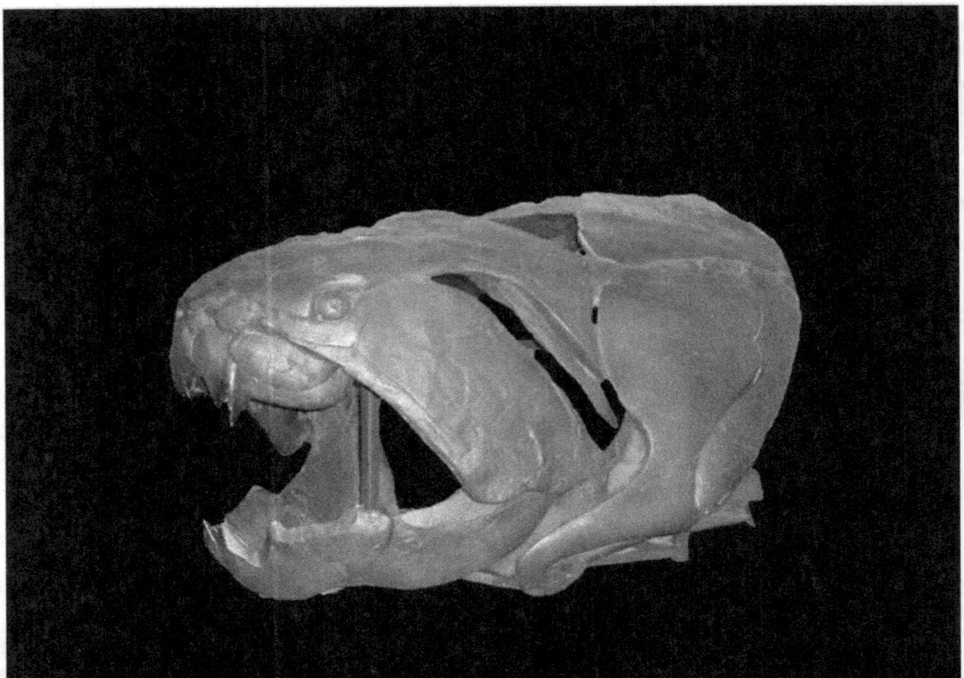

◄Dunkleosteus cast, Brisbane Museum- *Nicholas Kable*
◄Close-up of denticles on a Port Jackson shark, Point Cook, Victoria, Australia- *Philip Watson*

Buried inside the bramble shark's skin are specialised protrusions, denticles, which poke up like miniature spear heads. So rough is this effect on the skin that other smoother scaled fish like jacks will sometimes mob reef sharks and rub their parasite ridden bodies so roughly against the hunter that the shark can be forced to flee its harmless antagonisers. The armoury is not restricted to the skin. When the bramble shark opens its mouth covering the inside of its jaws are more denticles. These denticles are the first proper teeth. To enable them to work efficiently they are neatly ordered in rows. But unlike mammals there is more than one row of teeth in each jaw. Instead the rows of teeth extend backwards in sleek long lines, rather than running parallel to the jaw. The small sharp teeth are prone to blunting and falling out. This is something that the bramble shark is well prepared for. Being based on the same material as its skin the teeth are relatively easy to replace. The bramble shark's teeth are situated on a muscular conveyer belt. As old teeth wear down they are pushed out from behind and readily replaced.

Being covered in denticles also means that the body is considerably lighter and more agile than if it was covered in armour, a factor leading to the continuation of denticles in more modern shark species including the sea-urchin munching Port Jackson shark. The early sharks' spines were also considerably more enforced than those of the lancelet. The notochord became replaced by a rod of cartilage. As individuals mature the cartilage gradually hardens and becomes replaced with bone. Bones can support more muscle tone and generate more energy than the lamprey's notochord. It also allowed the ancestors of modern fish to compete with and finally out-compete monsters like Dunkleosteus.

There is another denticle covered fish that also sheds some light on the first fish. It is the most primitive living member of the sardine and anchovy family, the denticle herring. All of its more advanced relatives have replaced the denticles with softer smoother scales, which have an improved effect on streamlining. This ancient herring has another peculiar feature that has been lost by its relatives. Running along either side of its flank is a column of specialised cells that detect changes in water pressure. This is called a lateral line and allows the fish to sense the world in a way that can barely be comprehended let alone related to. Perhaps the closest way to understand it is to think of bringing two hands so close together that they don't touch but can feel the warmth radiating from each other. What the denticle herring feels is much more advanced than this. The lateral line feels the changes in the water around it as other objects and animals move around it. Above air this system would be useless but in water the slightest movement is carried straight past the lateral line, just as a splash in a small pool can send ripples to the other side. The advanced lateral line can quickly alert the denticle herring of either predator or prey. With the exception of most of the herring family, ironically enough, a lateral line occurs in basically all fish and the existence of one in the denticle herring indicates that once it was widespread in this family also.

The denticle herring can also smell advancing animals, prey or predators, far more efficiently than either the hagfish or lamprey. Instead of just one nostril smelling all the various odours around it the ancient herring has, like the bramble shark, a pair. With two nostrils it can compare what direction a scent is coming from and hone in on it. It is something that all backboned animals with jaws are able to do, both under the water and on land. This includes the bramble sharks. Bramble sharks, indeed all sharks also have lateral line. That they share so many characteristics with the denticle herring is barely surprising, they also share a common ancestor.

About 460 million years ago what had been a single family of fish distantly related to the hagfish and lamprey split in two. One group stayed in the ocean and became supreme hunters. We will return to them later. The second group, that which the denticle herring belonged to, ventured into fresh water. Under the pressure of stagnant water they developed swim bladders to assist breathing. This also enabled them to increase buoyancy control and move through the water easier. Competition applied pressure for them to become more streamlined to increase their speed. They reinforced their spinal columns and developed strong muscles to drive their fins. Some of these new fish used their fins to venture onto land, whereas others returned to their ancestral homeland and with their swim bladder and muscular fins conquered the ocean. Here they developed into species as distinct as eels, seahorses and tuna. These are the modern bony fish.

Like the spineless fish of millennia before some of these fish have to travel up river to breed. Atlantic salmon are some of the most dramatic to do, using their powerful fins to launch themselves out of the sea. In their passion they will face bears, eagles and even jump over waterfalls. But they do not do so alone. In July as they begin to mass in their millions lampreys are also assembled by the coast, feasting on any small prey that passes. One species in particular, the sea lamprey, has a particular interest in the salmon. But the salmon offer more to the lamprey than easy prey, they are ideal transport. Lampreys use their circular mouths to offer suction against the salmon's body. Their teeth grip into the muscles and allow the lamprey to hook themselves firmly into their target. From now on wherever the salmon goes the lamprey will follow.

As the salmon battles the torrents the lamprey continues rasping on it with its tongue, slowly feeding on the salmon from the inside. The parasite can suck 1.5 litres of blood and muscle from its host. Eventually the exhaustion will be enough to kill the salmon. But the lamprey still needs to travel forward. So it relaxes its mouth and easily unhooks from its dying victim. Salmon typically travel in vast shoals so it does not take long for the lamprey to find a replacement courier. It will continue in this way until it reaches its breeding grounds, in the process wearing down multiple salmon to an almost inevitable death. Lampreys may be thought of as primitive but this energy saving technique has been a crucial part of their spread throughout the ocean. They are also apt to take advantage of any new opportunities.

Recently the lampreys were offered the largest of such opportunities. The Great Lakes in North America were a crucial environment for many freshwater fish. The lakes were tantalizingly close to the sea but it was their isolation which led to their importance. Locked away from the ocean the local inhabitants had survived without the competition that had appeared in other shores. They were also protected from marine predators that could slip otherwise into the water. Great shoals of fish likewise sustained local fishermen who traded the fish for other essentials. The lakes also offered opportunities for transport to export the goods of the local region. But this transport would be limited while the lakes were still landlocked. So, early last century, a man-made link between the ocean and the Great Lakes was established.

Sea Lamprey attached to Chinook Salmon, Lake Huron, Ontario, Canada- *Sabine Atteln*

Briefly the local area thrived. Now the citizens could quickly and easily sell their fish without the additional cost of finding transport outside the lake. New marine species soon entered the lakes initially enriching the environment. But even the most productive of environments can only provide enough nutrients and food to maintain a limited number of individuals. So although there were more species each of the original species had a reduced population than before the lake was opened up. So they were more vulnerable, including to overfishing. Unlike their marine competitors the endemic fish did not have large numbers of relatives able to swim in and replace their numbers if disaster struck. So the original staples of the fishing community changed. That was when the lamprey came.

By 1954 the intruders had spread through all of Canada's Great Lakes and were breeding in nearly as many. Over several years they quickly decimated the local fish population. It is only after seventy years of relentless focus from local people and then the wider Canadian and United States communities and governments that the parasites numbers are gradually coming under control; and even that is only due to an extensive and expensive focus on them including targeted fish poisoning, releasing sterile males, regular trapping and, most ironically, using dams and weirs to restrict the lamprey's movement- basically replicating what the Great Lakes natural isolation was already doing before it was breached. By later 2024 some eight and a half million lamprey were being killed each year, and the population had finally been reduced by more than 90% of what it was in the 1950's, giving the surviving endemic species a real fighting chance again; primarily due to the poisoning which used a chemical that other more advanced fish could manage with but that the lamprey's body was incapable of processing. Yet it took only a few years to build the channel that connected the lake to the ocean.

It should not be forgotten however that that the sea lamprey did not enter the Great Lakes on their own. Their presence relied instead on the power of other bony fish to swim such great distances. Lampreys may eke out a living on the stagnant floor of rivers and lakes and hagfish may swarm over whale carcasses but it is the bony fish that truly dominate the marine world. It is only by literally clinging to the tails of more advanced species that these ancient jawless fish have been able to take advantage of the new canal.

But bony fish have serious competition for the claim of most successful fish group. There were those other descendants of the jawless fish swimming in those ancient oceans. Unlike bony fish these specialists stayed almost entirely in salt water. This meant that they avoided developing organs such as swim bladders to maintain their position and initially were restricted as bottom dwellers like the hagfish. Instead of swim bladders these predators instead later developed excessively oily livers which, with their natural buoyancy, helped to keep them mid water. They also found other ways to lighten their bodies. Their backbones began to stop development early so that they never hardened into solid bone. With only cartilage to support them these fish could not grow as muscular, certainly not muscular enough to support their bodies on land like other bony fish would later do, but they did have one advantage. They were considerably lighter. They may not be as numerous in species as the bony fish but they are superior and feared hunters. No other group patrols the seas as efficiently as the sharks and rays.

Chapter 2. Sharks and Rays

Frilled Shark, Canary Islands- ***Lin Ching Feng***

Sharks are the ultimate marine predators. One feature defines their success more than any other; their jaws. That of a great white shark are some of the strongest in the animal kingdom, capable of shearing right into the body of a two-tonne elephant seal. The jaws of the whale shark are some of the weakest of its group, yet still strong enough to drag a net full of fish back into the water, despite its reputation as being only a harmless plankton feeder.

The teeth are arranged in multiple rows in each jaw, in some species there may be up to seven rows. It is an ancient design and one that has been preserved in rocks dating back 400 million years. The bodies of sharks are normally too soft to be preserved as fossils but their teeth are different. The teeth are virtually the only hard part of an animal otherwise entirely made of cartilage. Some of these fossil teeth have provided astonishing insights into a long vanished world.

One style of tooth was, extraordinarily, shaped like a trident. Like a three pronged fork its main prong lay in the middle, surrounded by a sharp point either side. Such bizarre teeth while being able to catch prey would not have been able to effectively cut through muscle or sinew, limiting prey to only that which could be swallowed whole. But still teeth like this are dotted through ancient sediment indicating they must have been successful enough at the time. Eventually these forked teeth fade from the record until about 100 million years ago they disappear completely. Early scientists piecing together their knowledge of sharks accepted that they would never know more about the owners of these exotic teeth.

That was until the end of the nineteenth century when a strange beast was dragged up from the sea bed of Japan. Sliding along on the bottom of a fishing boat at first sight it resembled a deformed eel. Its thin tail slapped around like a ragged piece of rope, lacking the distinctive split lobes that characterise most fish tails. The brown fish struggling against the long line was completely unknown to the fishermen and its unique appearance did not assist identification. Its reptilian head seemed to be framed with a large fringe of skin that stretched down its neck. But it was what was inside that head that was truly remarkable. The mouth snapping at the fishermen was filled with those ancient forklike teeth. The frilled shark was the last survivor of a group from the Cretaceous, sixty six million years ago. But its form was almost identical to the few fossils of sharks known from about four hundred million years ago. The goggling eyes revealed something about how it may have survived so long. They were large and green, taking up most of the creature's head. This was a fish of the deep sea.

The depths where the frilled sharks live are so far from sunlight that they are freezing temperature; only the concentration of salt keeps it in liquid form. In this frigid water every morsel of energy is precious so the shark has to keep its movement to a minimum. The shark has a problem though. The only way it can breathe is through moving forward, forcing water into its mouth and straining out oxygen before it leaves its gills. It does so by sweeping gracefully through the water, looking even more eel like as it does so. The gums are filled with teeth but the trident shape doesn't hold prey well so the shark has to be selective on what it feeds.

The frilled shark may appear primitive but if she is female she may be carrying an ancient secret to their success. Deep inside may be developing not a clutch of eggs, but a live shark. Unlike most bony fish or jawless fish all sharks fertilise their eggs internally. Giving birth to live young may seem like an advantage to every animal that can achieve it but it has its drawbacks. Firstly eggs and sperm must meet while still inside the parent's body. This requires the development of an organ in the male that will enable him to transfer his contribution to the female. Dangling behind the male like a pair of sausages are organs specialised for this very purpose.

When people first began studying the copulating methods of sharks they thought that the male used these extensions to cling onto the female and called them claspers. The claspers do not actually clasp but instead insert inside the female's vent and begin to inflate with water. As they expand the claspers begin to release thousands of sperm into the female's vent, all of them competing for a single egg. After much wriggling around the male's job is done and he leaves. The riskiest part of a frilled shark's reproductive life is concluded, for him. For the female the most hazardous part of her life is about to commence.

Nurturing offspring internally puts the mother at risk of starvation as her body has to constantly feed the young, despite her own limited food supply. To avoid that, when placed under increased stress the shark may abort. Via sacrificing its offspring the frilled shark will have the chance to reproduce again under less hazardous conditions but it has also wasted up to three years of its life.

For the frilled shark the challenges of breeding and even general existence are harsher than most. Living at the ocean depths food is hard to find, and the frilled shark may not feed for months at a time. The icy temperatures now work to its advantage, as the shark is cold-blooded she responds to the low temperature by keeping her metabolism equally slow.

This also slows down the development of the infant significantly. This can be seen most effectively by comparing it to another group of animals that invest in live birth, mammals. On land the elephant has a pregnancy of twenty one months, the longest of any mammal. The sheer size that the elephant calf has to grow to before it is born has a significant effect on the length of pregnancy but this is only half the reason. Whales give birth to young that are even larger but even the biggest of them only has a pregnancy of eleven months. The reason that the elephant holds onto her calf so long is so that it enables her young to be more advanced at birth, in this case not just physically but cognitively. Like the frilled shark the elephant usually carries only one offspring and the mammal invests her energy significantly into the survival of her calf.

Frilled Shark baby still attached to its egg, Honshū, Japan- *Yurimaoh*

Birth does not mark the end of the elephant's investment in her young. Now the mother's body starts to produce milk to nourish the calf over the next several months. She will also need to teach it how to feed and what to eat. Over the same amount of time the frilled shark continues nurturing her infant inside her body, hunting down prey in the depths of the ocean. When she finally gives birth she will have carried her offspring for three and a half years, one of the longest pregnancies in the animal kingdom. Her baby is already capable of swimming for itself.

It has been inside its mother so long that it is basically a miniature adult. When it gapes its mouth is already full of teeth and it swims nimbly after prey. The mother's investment has paid off. Birth marks the end of any need the newborn has for either parent. It heads into the darkness alone. If it ever sees its mother again it will view her as either competition or a potential mate.

Despite its advanced birth something has kept the dominion of the frilled sharks restricted to the frozen depths. If it rises too near to the surface it becomes an obvious target for larger more agile predators, modern sharks. There is one simple difference between the two groups; the number of gills. The frilled shark has six gills; nearly all other sharks have only five.

There are two other ancient sharks with six gills and another even more distant species with seven gills but like their cousin these have all been restricted to the depths. These species are generally considered primitive, barely changed in one hundred million years but they are still significantly advanced when compared to the frilled shark's frail physique. The first major change is at the tail. Rather than a single lobe, the tail is divided into two parts like a more typical fish tail with the top section significantly overhanging the lower lobe, giving it that much more thrust and power when it is swimming.

The six and seven gilled sharks' mouths are also different. The teeth vary between species but none of them have the trident design, instead some of them are shaped like vicious fish hooks with an additional sharp point midway down to inflect extra damage, while others look like they have been moulded off miniature saws. Like the frilled shark they have multiple rows of teeth arranged conveyer belt style but only the first row is used for eating, the other teeth being held in reserve so that as each primary teeth falls out a replacement is ready to slot into place. This is the same dental arrangement of all sharks but despite all these advancements these sharks are still obviously primitive in one other respect. The modern species of shark that patrol the surface of the ocean, indeed nearly all of the ocean, have only five gills making them more oxygen efficient. Better breathing allowed greater bursts of speed. It also encouraged streamlining.

Modern sharks have morphed from the eel shape of their ancestors into pinnacles of muscles driving through the water. Muscle is heavy so to counter this most species have rich oily buoyant livers to help lift the sharks out of the depths. Some species have developed massive livers as a result, making up to a quarter of the animal's weight. But in cooler temperatures most modern sharks can't find enough food to fuel their active lifestyle. This is why the ancient sharks have been able to survive at their slower pace in the frigid ocean depths. Indeed six gilled sharks can not only live here but reach giant proportions of five metre lengths. Being so large and energy efficient the six gilled shark can survive for months without eating. When they need to they can show the perseverance to catch a seal or another shark straying on their territory. But it is the modern five gilled sharks that are the true success story of this group of predators.

The epaulette shark is streamlined compared to the frilled shark but still has a fairly sluggish shape when compared to species like the sleek blue shark. This is because the epaulette uses its flexible body to squeeze into tight crevices in its endless search for food off the coasts of Papua New Guinea and northern Australia. The shark is about a metre long with a blonde body marked with small black spots. Lying prominently above each pectoral fin is a particularly large spot bordered by a bright white outline. The epaulette spends most of its time sucking in crustaceans and starfish hidden in the substrate with its vacuum like mouth, then crushing its catch between its teeth. To do this is has developed another technique to assist with breathing.

Many sharks have to swim forward continuously in order to create enough pressure to force water into the mouth and out its gills. Were they to bury their head beneath the soil searching for food they would quickly suffocate. But the epaulette has two large holes beneath its eyes called spiracles that are used to suck extra water into the shark's mouth, pumping it back over the gills in the process, allowing the shark to breathe independently of its swimming. This allows the epaulette to rest on the sea floor and dig for hiding prey.

Epaulettes have a particular preference for hunting in the shallows around a reef. This can be a particularly volatile location for not only can small clams and ghost crabs be trapped by the exiting tide so can the epaulette shark. Depending on the size of the rock pool it may clear out any suitable food in a matter of minutes. At night being stranded in one of these pools is only a temporary inconvenience but in the heat of day it is another matter entirely. In summer it may reach over 40°C in the shade and the little oases soon begin to evaporate. The spine of the shark soon appears above the water followed shortly by its flanks. Its flexible body presses as deep into the pool as it can. Even this may not be enough to keep its entire body moist. The epaulette has one last survival strategy. It hurls itself out of the water and begins to drag itself over the baking sand. Surprisingly, unlike any other shark, it can breathe out of water. It cannot do this indefinitely but it will allow it to stay conscious long enough to reach another pool. By travelling from one isolated pool to another the epaulette can systematically harvest its stranded prey without its food in turn being able to escape until the tide returns. So with its unique gift this species can survive on the edge without any competition.

The epaulette is a relatively recently evolved species but on the other side of the world are sharks whose family have been feeding on invertebrates since the Jurassic. Off the Eastern coast of Southern California down to Peru are found horn sharks. Like the epaulette they have spiracles to allow them to breathe while they dig into the sand at night but they also have peculiar bulldog shaped heads. Sharp and angular teeth line the front of its jaws. These are unusually short allowing them to slice through their prey easily before swallowing it. Such a dental design is ideal for worms and other soft prey but the horn shark also targets sea urchins covered in dense spines and snails grazing on the kelp. Small sharp teeth would barely be able to scratch such armoury. So at the back of its mouth are large flattened molars that can be used to grind continually at whatever defences it encounters until it can reach the juicy morsel inside.

The smell of ripping flesh can quickly attract other sharks, some of which would find the horned shark an appetizing prey as any sea urchin. But the Californian horned shark has its own protection. At the front of both of its dorsal fins is found a short spike that can easily pierce through skin. The fact that it needs such weapons is a vivid reminder that this shark is only small compared to others of its group and is prone to fall victim to them. Its streamlined shape makes it easier to escape but it also allows it to undergo more intense and prolonged swimming. In Sydney a different horn shark, the Port Jackson shark, uses this practical body shape on its annual migration, of up to 800 kilometres, along the coast to breed.

Like all sharks the eggs are fertilized internally but unlike the frilled shark these eggs are laid well before the infant hatches out of them. Horn shark eggs are shaped like a corkscrew. In the Californian horn sharks this shape alone is enough to allow the egg to be dug into the sand by the strong current.

Crested Hornshark eating an egg of a Port Jackson shark, Botany Bay, Sydney- ***Harry Rosenthal***
A surviving juvenile Port Jackson shark, about 6 weeks old, Mornington Peninsula, Victoria, Australia- ***Philip Watson***

Port Jackson mothers go to an extra effort for their young. Seizing an egg in her mouth she gently prises it into the ground, usually in a cave or crevice which she has used before. The mother may repeat this process sixteen times. But not all her eggs are the same. Only some of the Port Jackson's eggs have been fertilized and these are laid in the most secluded areas. The others are sterile duds and any predator that eats them may be satisfied enough to miss the real eggs. Their extra vigilance is needed. Wobbegongs are carpet sharks, the same family that the epaulette belongs to, that live in the area year round. Like hagfish they are seasonal feeders, glutting themselves on eggs and the developing young. Other sharks, like the closely related crested horn shark, are also interested.

Sharks have one advantage when it comes to finding food; they can detect electrical impulses. The specialist equipment that allows them to do this is called the ampullae of Lorenzini. These are scattered evenly around their face like five o'clock shadow and allow the shark to detect the miniature amount of electricity being used each time a fish moves a muscle, breathes or even beats its heart. A hunting shark can easily discover the developing juveniles, even while still hidden in their eggs buried beneath the sand.

The baby Port Jackson sharks are not completely defenceless however. Multiple clutches hatch simultaneously releasing hundreds of infants into the open. It is an ideal opportunity for wobbegongs but as they focus on one easy meal others inevitably make their escape. Few hatchlings will survive to adulthood but enough will that in future years some will return to lay their own eggs. But other sharks develop in places where few egg predators can survive.

Northern Europe in December and the seawaters are close to freezing. Most of the fish that shoal here in winter have long since left for warmer waters. So have most of the adult sharks but not necessarily their young. Like the Port Jacksons swell sharks also develop in eggs. Unlike the sunny coasts of Australia few hunters remain in winter so the egg can be left unprotected. The eggs appear so mysterious that when they were found washed up on the shore they were called mermaid's purses, a name that persists to this day. Tendrils projecting from the egg hold it tightly against a piece of kelp so that it is unlikely to be separated even in the worst winter storms. The dense protective layer of the egg also insulates the embryo from the frozen waters around it. Inside the egg an umbilical cord to the yolk will provide the nourishment for the developing shark over the next eight months, until it uses the two spiked rows on its back to cut its way out of the egg.

Adult swell sharks can not rely on the security of a protective egg case for their security. Without defensive spines in front of their fins or toxic poison they would be an easy meal for any predator large enough to attack it. So the swell shark uses its flexible body to its advantage. A fully grown swell shark is about 1.2 metres, large enough to be out of limits for most dogfish, a larger species of shark, but not big enough to take them off the menu completely. So when the swell shark feels threatened it inhales rapidly. Water floods its body and it starts to expand like a balloon at a town fair. Having no spines defending its body means that there is no risk of impaling itself as it expands. There are few animals that will take on the formidable task of attacking a fully swollen shark.

The adult swell shark may be relatively predator free but it is not immune to old age. It will be lucky to live half a decade. The spiny dogfish that patrol the same waters on the other hand may live up seventy. They can grow to two metres or so. Further north live dogfish that are three times the size of their European relatives. Greenland sharks are the largest dogfish in the world. Some approach six metres, threatening the claim of the great white as the biggest carnivorous shark. Their movement is so relaxed that they often have invertebrates growing from their skin; small crustaceans may even cloud over their eyes. Not without reason are they also called Sleeper sharks. One female's eyes were cabon dated to be 390 years old, with a margin of 120 years, meaning she could potentially have been anywhere between 200 and 500 years of age. With their sedate lives they mange to have pregnancies that are estimated to last between eight and eighteen years. Yet, still these individuals often look healthy. Caribou have been found amongst their stomach remains, presumably scavenged from a sunken carcass. But on occasion Greenland sharks are much more active than their usual slovenly behaviour would indicate, travelling several kilometres below the water in search of giant squid. This is no ordinary meal. Not only is it enormous, up to eighteen metres long, it is formidably armoured. Its barbed arms are fully capable of inflicting dangerous wounds. Large eyes allow the squid to pick up the faintest movement in the darkness and be ready for it and its beak is as short and sharp as a parrot's. In the Arctic these sleeper sharks are the only predators of adult giant squid, but in the darkness it is still not known precisely how they achieve this.

One dogfish, the pygmy shark, has its own particular problems. It is considerably smaller than the Greenland shark, or any other shark for that matter, barely 25 centimetres long in the largest individuals. The pygmy has the distinctive strong teeth of all modern sharks for defence, but at such a size most predators could swallow it without being bitten. So it has a spike embedded in front of its dorsal fin that can be erected to catch against a predator's mouth. After that it is usually spat out. It has another, more discreet defence. In the deep waters where the pygmy lives the slightest silhouette could indicate its presence to predators below. Its underbelly is illuminated by large light reflecting organs so that any hunters below won't be able to distinguish the shark from the sunlit water above and will ignore it.

Such small sharks are at real risk of being not predator but prey. The cookie cutter shark is not much larger than the pygmy and also uses light reflecting organs for camouflage. The organs do not cover all of its body. A small patch of the cookie cutter's skin behind its neck is left bare. From a distance the resulting black spot resembles a tiny fish. With any luck a juvenile sea trout or other modest hunter will be drawn in. As the trout moves in to attack the visual illusion the cookie cutter simply twists his neck around and swallows it. But larger predators may also be attracted. Dolphins have a particular vulnerability for sampling these sharks. But dolphins are too large for cookie cutters to kill. Indeed dolphins often swallow fish of a similar size whole. But this enterprising dogfish has managed to turn danger into an opportunity. As the dolphin hones in on it the shark turns on the cetacean. The cookie cutter has a curious arrangement of triangular teeth that protrude out its jaws. It uses its mouth like a tin opener, neatly carving a circle of flesh out of the dolphin before escaping into the darkness. The cookie cutter has another use for its teeth. Like all sharks its teeth regularly fall out and are replaced. In other species they simply drop to the ocean floor but the cookie cutter does not let anything go to waste. When a tooth comes loose it is swallowed and sucked into the shark's gut. As the teeth break down the cookie cutter extracts all the calcium out of them.

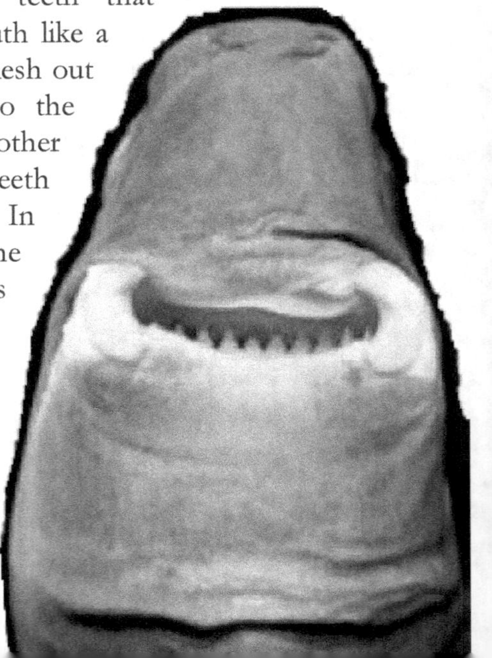

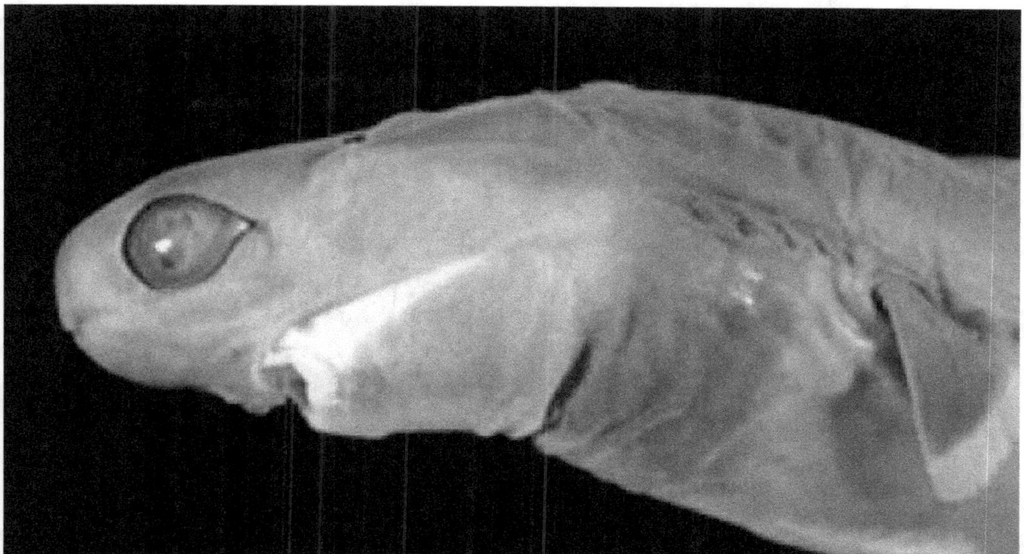

◄▲Cookie cutter shark- *Karsten Hartel, Marine Fisheries Review*
Requiem Shark feeding in sardine bait ball, South Africa- *Pcowell*

Due to this process unlike in any other shark or ray the cookie cutter does not have a skeleton of cartilage. Instead, most of its skeleton is comprised of solid bone like bony fish and land vertebrates and consequently is much firmer and stronger.

The cookie cutter may be a ruthless hunter but its sinuous shape provides little alarm to any people that come across it. Nor could the wobbegong or dogfish be said to inspire scenes of mad panic even though they can bite viciously if threatened. Requiem sharks on the other hand have an almost undeserved reputation as mindless voracious predators. The majority of species are as passive to man as any dogfish. But not all. Of the five reputed 'man-eater' sharks four belong to this family; the oceanic white tip, tiger, greater hammerhead and the bullshark. As well as speed and an insatiable appetite they have another adaptation to aid their hunting. They have a nictitating eyelid, a third transparent eyelid that slides over the eye. During hunting this eyelid provides extra protection from any blood or debris released into the water around it. All are undoubtedly efficient predators but the behaviour that produces this is anything but mindless.

Along the South coast of Africa bronze whalers congregate to hunt sardines on their annual migration. A change of current is all it takes for the shoal to be pinned against the coastline. Bronze whalers can take their pickings of the healthiest, fattest prey. The smaller fish are entirely helpless to move until the cool current they rely on continues to push its way up the coast, and away from the predators. The whalers gorge themselves while they've got the chance, their progress can be seen be as great gaping holes in the otherwise near endless shoals of sardines.

When the sardines are further out from the coastline the whalers rely on other tactics to corral their prey. The sharks come up from beneath the sardines and surround them. The circling sharks will even join forces with dolphins, seals and gannets to group their prey together. It is a myth, and a dangerous one, that where there are dolphins there are not sharks; in an aquarium setting dolphins certainly can attack and even eject sharks out of the water, but here in the wild they need each other. While both the mammals and birds need to surface to breathe air the sharks can maintain the attack from below. Circling around the sardines the whalers slice in amongst the massive group until the fish becomes so disorientated that they easily blunder into their teeth.

Other sharks are drawn in by all the commotion. Grey nurse, spinner sharks and even the mighty great whites all converge on the shoals. Spinner sharks in particular have one of the most dramatic feeding techniques, circling in an increasingly tight circumference as they ascend towards the sardines until they plunge out of the waves, fish in mouth. Another skilled hunter, the tiger shark, is also not averse to sardines but then there is not much they won't eat. As well as fish they've been known to consume turtles, seabirds, abattoir waste and even car tyres. Most bizarrely, a three metre long individual on the Great Barrier Reef vomited up an echidna, the land living egg laying spiky mammal that has been known to swim between local islands; clearly those spikes proved more trouble than the bite-size creature was worth. Annual events, like the sardine migration, provide a welcome reliable nutritious meal as the tiger sharks appear, seemingly out of nowhere, from deeper water. But the tiger sharks' wandering is far from random and like several other species of oceanic sharks they embark on the same migration annually.

As infants tiger sharks are covered in thick brown stripes to help them blend in among the mangroves where they were born. But as they grow and begin to hunt in the open water the stipes slowly fade until they dissolve into little more than streaks. Sometimes the adults swim up river into more brackish water, where the majority of its competitors are unable to enter. It does so at a cost however. Fresh water, by its very nature, is absent of the minerals that marine fish need to maintain their body processes. It also neutralises the shark's salt content, slowly absorbing it out of the blood stream. After a certain point the tiger sharks simply have to return to the sea.

But the tiger shark has a relative that regularly enters completely fresh water. Until the 1970s bull sharks used to swim so far into the Amazon and in such numbers that there was an entire fisheries industry targeting them, until they were overfished. This super river is populated by thousands of small fish, so diverse that over a thousand may be unknown to science. Most of these fish are tiny, barely a snack for a 230 kilogram giant. Occasionally the carcass of a medium sized mammal like a tapir or a capybara, the largest rodent in the world, may float down the river. In the tightly packed South American jungles large animals are sparse and the chances of one washing into the river are even rarer. So bull sharks here are forced for the most part to continually hunt the bite sized fish that shoal around them to survive.

Bull sharks occur throughout the world, patrolling coral reefs off Indonesia, the coast of Florida, swamps around Bangladesh and even travelling into fresh water in the Mississippi and Russia. Their most dangerous quarry however is found in Africa. Like their South American relatives the bulldog sharks here travel a thousand kilometres or more inland using mighty rivers like the Zambezi. There are jungles here too, but most of the bull sharks live in wetlands that cross the savannah. The plains mammals are typically large so they can migrate following the rains, often in great numbers, some like the wildebeest in their millions. Unlike in the Amazon water is comparatively scarce so many animals congregate at the same waterhole to drink. The substantial herds can often be heard long before they arrive and the thudding of a thousand hooves through the water sounds as a clear dinner bell to predators. Bull sharks, like all fish, cannot hear out of water but when the antelopes' muzzles start slurping they can easily pick up the sound. Their lateral line also helps them pinpoint their quarry.

Bull Sharks, Florida- ***Sam Riley***

Bull sharks are not the only predators here. Crocodiles also gather in anticipation of the herds. A crocodile can eat a young bull shark, ripping it apart with its formidable mouth. Crocodiles are impressively encased with solid scales that act like a suit of armour. Adult bull sharks are capable of being equally dangerous as they can seize their enemy by the spine and carve them open with the skill of a butcher. So both predators tend to treat each other with a mutual respect when there is plentiful prey.

There is another highly aggressive creature in these waters, the hippopotamus. It may be a herbivore but it is also the most dangerous animal in Africa, killing more people each year than elephants, lions, buffalo or sharks. At one and a half tonnes it can trample over anything that tries to threaten it. The head can rip through grass or an attacking lion like a bulldozer. Protruding from the jaw are two immense canines that serve no purpose in feeding. Instead they are used in territorial disputes, slicing into one another. These canines can also be used to decapitate an attacker. It is no surprise that few animals attempt an attack on a hippo. Except a bull shark.

The ideal opportunity is when the hippos give birth. The pregnant animals normally isolate themselves from the group. A newborn would be at risk in a herd which, with its quarrelling nature, might accidentally trample it to death. The hippo may choose to give birth just outside the water but it is equally likely that she will deliver her offspring in the water. A release of blood from the birth canal into the water signals that the mother is ready. By standing at the edge of the lake in deep strands of papyrus the mother blocks most of the smell spreading across the water and drawing in predators. But bull sharks with their highly sensitive sense of smell only need to smell a few particles of blood mixed in the water to be drawn near. Most of the time they will ignore this highly seductive dinner bell. But if the migration of the antelope has taken longer to arrive than usual they may choose to attack.

A hippo calf can swim within five minutes of its birth. It can also run onto dry land where no shark can reach it. So the shark must quickly take advantage of its opportunity. The mother hippopotamus keeps her baby close to her deadly jaws. But she is exhausted from the effort of delivery. The bull shark slips in and rips a mouthful of flesh out of the calf, testing the waters so to speak. As it darts in the shark's transparent nictitating eyelid slides forward so that it can continue to see even as the blood flows over its face. The shark begins to drag the calf deeper into the river; if it can drown it the shark will have a meal that will satisfy it for several months ahead. Sometimes hunger will even drive them to attack an adult hippo.

A large kill often attracts another shark. Armed with teeth and sheer power a shark can be deadly not just to its prey but competitors of the same species. Any injury to either could prove fatal so sharks have had to become highly efficient communicators. Some bony fish use flashes of colour to indicate emotion much like a chameleon does, or how our own cheeks can flush red when angry or embarrassed. Most sharks are colour blind so they have had to create their own style of communication. An oceanic white tip feeding on a whale carcass can rely on its size alone to deter most competitors. But occasionally other giants arrive which are more than a match for it, a great white for one. It uses one of the simplest and most effective threat signals; opening its teeth crammed mouth in an obvious intimidation display. If any shark could match a lion's roar as a territorial display this would be it. A great white might not have a voice box but the sound of it crunching its enormous teeth together as it concludes the display is a spine tingling affair. Such a show can be used to quickly drive a smaller rival away from a kill. Survival for an oceanic white tip comes down to being able to accurately read the intentions of other sharks and avoid being attacked from a misunderstanding.

On the reef sharks are not only smaller but also encounter each other more frequently. So, one of the sharks here has developed a form of conversation with its allies that is a little more subtle than simply flashing your teeth. The grey reef shark is a typically streamlined coastal shark that patrols the coral in flexible groups. It has developed some of the most beautiful and choreographed ways to communicate by using its whole body to send out complex messages. When two sharks come across, for instance, a sinking carcass of a dolphin mistakenly caught as bycatch, they begin posturing; stretching out their fins and tail to exaggerate their size. If this is not effective they begin to contort their spine, pushing their backs up until their dorsal fins loom high above their head and curling their tail down, taking up the maximum area their body will allow them. If neither feels intimidated they begin circling around, testing for any weakness or stiffness of motion. Only if neither shark backs down will they fight and even this will likely be brief. Even a short confrontation is rare due to the effectiveness of the communication between grey reef sharks; it is more efficient to find another meal rather than fight and potentially be injured by a larger shark. Indeed, about the only time that a reef shark attacks anything that isn't food is when a curious human blocks it against the current leaving it with no alternative to physically defend itself in order to escape. In these circumstances the shark's bite may certainly be painful but not fatal and it allows the shark to get away and return to hunting prey small enough for that it can actually eat. Many such bites could be prevented if people understood the sophisticated predators' language as competently as other sharks do and simply retreated.

Reef sharks use other signals to distinguish each other; especially when it comes to telling what species they belong to. Reef sharks are almost colour blind so most species use broad bands at the tips of dorsal fins to distinguish each other. Grey reef sharks are the exception that proves the rule, being an entirely bland grey colour including on the fins but they are the only reef shark to use this colouration. Black tip reef sharks and white tip reef sharks are both well named for the distinctive colour marking their fins.

White tips not only recognise each other but stick together, principally at night when they hunt. The sharks' electric sensors give them a distinct advantage over their prey but it is their pack mentality that truly grants them success in making a kill. They are solitary predators, but the compact design of the reefs leads to several sharks combing the same area, squeezing in and out of narrow nooks in the coral and fighting over the carcasses of unearthed parrotfish.

Despite their complex communication techniques the reef sharks occasionally bite each other during the excitement. White tips have protective measures against this. Like most sharks they have a layer of denticles, dense protective scales, running across their body. When moving forward the scales are forced back smoothly and keep the shark streamlined as it pushes through the water. Any accidental bite is likewise deflected by the shark's skin. Females have a special need for this protection. During mating the male grasps onto the female with his mouth so that he can maintain contact with her. Even the gentlest of white tip mouthing is capable of ripping through the toughest skin. The scales of a white tip are formidable but some females still get long wounds running across their bodies. It is only due to their skin that they can survive the hardships of courtship at all.

White tip Reef Sharks, Revillagigedo Islands, Colima, Mexico- *Barna Takats*

As with their hunting techniques white tips mate mainly at night; when more efficient predators are resting. The presence of larger diurnal hunters on the reef means that white tips tend to sleep in caves during the day. Once they've found shelter they rest on the sand, pumping water into their gills through small spiracles behind their eyes. When groups form the white tips show what could be considered their tender side; slumbering gently on top of each other.

In the Caribbean another family of sharks, the hammerheads, has also been forced into socializing. In the day they hunt for shellfish and sea urchins hiding under the sand. The head is stretched out like a medieval mallet and seems a great incumbent to the animal but it comes with a purpose. Like all sharks hammerheads have those specialized cells, ampullae of Lorenzini, running across their face to detect electrical signals but the distorted shape of the head means that these cells can be spread over a larger area. The hammerhead uses its head like a metal detector, waving it back and forth in search of hidden prey, especially its favourite meal, stingray. Some hammerheads like the greater hammerhead are massive and occasionally hunt humans as well as each other. Yet annually they overpower their basic aggressive instincts in spectacular form. Off the coast of Costa Rica hundreds of greater hammerheads converge. As they reach their climax packs several thousand strong cling together. Then the giants begin to peter off and disappear into the deeper water as mysteriously as they arrived. They will not see each other again for another year.

A smaller species, the scalloped hammerhead, comes together in large groups not just yearly but daily. As the sun rises they congregate not in hundreds but thousands. They align over each other in dense masses. It may seem as if this was some elaborate courtship or hunting behaviour but the sharks neither mate nor feed while in these vast groups. Indeed all of the group are females and will use force to keep it that way.

The hammerheads with their sensitive heads have low vulnerability at night but during the day they lack the speed to catch most fish or to escape larger predators, like false killer whales, that prefer to hunt while they can see their targets. Sharks are not usually depicted as creatures sheltering together from a threat and larger individuals certainly seem to resent such socialising. A female frustrated by the intrusion into her solitude will spiral and head-butt the hammerhead beneath her. Such expressions of vexation occur throughout the day but for the most part it is as far as the conflict inside the group goes.

Males avoid such social stresses by heading further out to sea during the day but should they become discovered by a passing pod of toothed whales they only have their size to defend themselves with. As the sun sets so the fish the sharks prey on begin to return to their nightly shelters and the females hammerheads spread out across the reef in pursuit. Now these sharks show the characteristics they are more known for; they emerge into the night solitary and graceful hunters, with little tolerance for competition or companionship.

Other hammerheads also form small shoals, though less spectacularly than the scalloped species. The bonnethead regularly moves around in groups of half a dozen. As they move around in the seagrass swallowing the occasional squid that gets scared into the open they also feed on something entirely unexpected, the seagrass itself. The sharp blades of their teeth nip through the plants with ease although their intestines are not as adapted to this vegetarian element of the diet indicating that this behaviour is a comparatively new development. The unexpected fare makes up at least half of their diet, scientists studying them in captivity have been able to feed bonnetheads a diet consisting of 90% seagrass and the rest squid and yet the sharks still put on weight.

◄Scalloped hammerhead sharks shoal, Galapagos- *Alex Vog*
◄Greater hammerhead shark, Alice-town, Bimini, Bahamas- *Image-Source*

Behaving as a herbivore for much of its life, and feeding in the open to do so, does make this species more vulnerable to other sharks with less catholic diets, which is part of the reason why they stay in small groups as much as possible.

Even with this well developed social tendency on rare occasions a female bonnethead may become separated from the group during the crucial time when she becomes fertile. Should she be unable to find a male her eggs are able to develop without insemination by a process called pathogenesis, or virgin birth. In the wild there are few occasions when a bonnethead will become so isolated but in an aquarium often a bonnethead will be kept alone or with only a few other females. In these conditions the self-cloning sharks are still able to reproduce and ensure their genetic survival, this time via their offspring being, unknown to the mother, potentially shipped around the world.

Further out to sea there is a shark that is social not just at rest but even while hunting. Common thresher sharks are large, more than five metres long, but this is not their most distinguishing feature. Thier skin has violet undertones across most of the back fading to a light blue on the abdomen. In the ocean these colours are hard for people to distinguish but most fish have better colour vision than mammals, including the baitfish that threshers commonly prey on. What a terrifying vision it must be for them too. The thresher has a surprisingly flexible body for a shark that now begins to curl in towards its quarry. Trailing behind is a large fin, stretching out more than a third of its total length. As the baitfish attempt to escape they instinctively flee the mouth and try to sneak past the tail. But this could be their last mistake. The thresher does not have the longest tail per body size for nothing. It sweeps it tail back briefly then cracks it like a bullwhip through the helpless fish. The strength being utilized to move the tail is both formidable and deadly. A fisherman was allegedly decapitated after a caught shark sliced his head off. This is a story that seems to have been more legend than fact; there are, to date, no records of the thresher shark killing anyone; nor is the tail sharp enough for such an occurrence to be likely.

It is not the only legend associated with these skilled hunters; they were previously accused of teaming up with swordfish to hunt whales- using their tails to smash into the side of the whales; as their small teeth would hardly be able to break through a whale's tough skin let alone its flesh this must be considered even less likely. There is no doubting the efficiency of its hunting technique when used against a shoal of baitfish however. The shark frequently combines this with another technique, team work.

As a pair of thresher sharks draws close to their quarry one attacks the school of fish head on while its companion snakes along to the other side. Together they cut in between their prey. Occasionally another couple of threshers will join the hunt. In order to eat safely the sharks have to control their natural urges to dive into the food. Instead they surround the shoal with both their numbers and their elongated bodies, flicking individual fish out with their tails and swallowing them whole. By hunting together the threshers can easily surround and catch their prey.

The team work involved in the hunting tactics of these sharks is surprising complicated. But what led the threshers to this strategy is even more remarkable. Thresher sharks belong to a family of sharks, the mackerel sharks, which are among the most advanced animals in the ocean. Fish are typically thought of as cold blooded. For some groups like herring this is basically true, they derive the heat they need to run their body almost entirely from the surrounding water temperature. Consequently as the ocean cools mildly during the night the activity of these fish is similarly reduced. The use of muscles while swimming does produce heat as a by-product although the herring quickly loses most of this in the cool water. But mackerel sharks are different.

Mackerel sharks have a thick layer of muscle surrounding their body to trap internal heat. This allows them to prevent internal organs like the brain from being chilled during the night, allowing to be more complicated as a result. But this protective muscle layer comes at a cost; it prevents the sharks almost entirely from absorbing heat from the surrounding water. The sharks maintain their body temperature in the same way as the whales and seals that share their world do. They produce their own heat; unlike any other shark group, and the majority of fish, they are partially warm blooded. In the open ocean this provides the extra amount of energy to pursue and take down their prey. Extra energy also requires additional fuel. Other sharks can wait several months between kills but mackerel sharks must hunt successfully at least every few weeks to survive. It is this constant risk of starvation that has moulded threshers into such successful hunters.

Pelagic Thresher Shark, Philippines- *Riley Wyna*

The thresher may grow to a great size but even so most of this consists of its slender tail, reducing both the shark's weight and its overall energy needs. The salmon shark in comparison has quite a stocky build including the tail and weighs a decent two hundred kilograms despite only being three metres long. With its powerful build it launches itself after prey like a torpedo, quickly outpacing the small fish that form the basis of its diet. Salmon in particular have one highly desirable property; their bodies are filled with oxygen rich muscles. This species needs all the fuel it can get; its body can be up 21 °C higher than the surrounding sub-arctic conditions. But even the salmon is not high energy enough to completely fuel the shark's energy. Only other warm blooded animals are capable of sustaining it. Further out to sea tuna also keep their bodies heated in a way similar to mackerel sharks. But tuna are out of reach of the salmon shark for most of the year. So the hunter makes a welcome snack of any unwary sea otters for an extra energy boost. The salmon shark has an even larger relative that has a hunger so intense it lives almost entirely on warm blooded mammals. It is known worldwide as a man-eater, the great white shark.

Giants of this species have been measured at more than six and a half metres long. The largest Great White ever caught was 7.1 metres, though this was extreme and there is much debate as to what the regular maximum size of these fish actually is. To keep these immense boilers burning requires something almost unique in the animal kingdom, intelligence. Across the temperate parts of the world seasonal changes alert animals it is time to breed; changes like shifting currents, full moons and rising temperatures. It is understandable for animals to know the cues for their own breeding but for great whites this is not enough. They also have to be aware of the triggers of great breeding congregations for all their prey. Each year great whites undertake regular treks in search of their next feast. It used to be believed they would navigate by the sky to find their destination but most of the time the sharks swim too deeply for this to be plausible. Instead they use their senses, already tuned to electrical sensitivity, to pick up one constant geographical marker- the gravitational pull of the poles. This way they can navigate successfully to the same destinations using the constant signals coming from beneath the earth's magnetic surface. This allows them to not only survive on seasonal blooms across the planet but to locate these areas again.

In South Africa changing currents lead to vast shoals of sardines, in West Pacific the summer causes vast blooms of algae and herring and off the coasts of Australia, until comparatively recently, great heaps of offal dumped in sewerage led to huge congregations of albatross. In turn these featured as regular pit stops for great whites. Individuals have been found to make annual migrations but because their own population is limited due to being a super predator the number undertaking these journeys was not large enough to draw attention for many years. They are now known to undertake some of the longest journeys of any shark; one restless male was able to swim a distance of 10,000 km in 150 days. In the Farallon Islands, about 50km to the west of California's Golden Gate bridge, one group of great whites begin their travels south east to Hawaiian fishing grounds for winter. But their time here is only temporary. As the seasons change and Californian waters begin to warm once more the great whites return to the isolated islands and the spawning shoals of fish and the furry mouthfuls feeding amongst the shoals.

The Farallon Islands play host to the breeding grounds of thousands of seals in spring; largest among them the elephant seals, formidable hunters in their own right. Their joyful jumping and squeaks as they chase after herring echo easily under the water. Such a sound acts as a dinner gong for the most feared shark in the ocean and it hasn't come for the fish. The great white tracks the seal from such a depth that the mammal is unable to detect it. Once the great white sinks below a few hundred metres its powerful eyes are unable to see the seal but its distinctive lateral line can still sense it. Its internal ears allow it to pinpoint almost exactly the coordinates of the seals above. The shark draws nearer then, before the seal even has a chance to notice, begins to shoot up towards its naïve victim. As it accelerates it begins to reach speeds of forty kilometres an hour. Unlike requiem sharks the mackerel sharks do not have an additional eyelid, leaving their eyes at risk from infecting blood and the defensive teeth of victims, so it rolls them away safely back into its forehead at the moment of attack. It is effectively hunting blind, relying on the combination of its other highly tuned senses to make a kill. It bursts onto the surface with such speed that its whole body can emerge out of the water, often with the seal grasped inside its mouth. Even if the seal manages to avoid its hunter's jaws the force of the giant crashing back into the water on top of it will severely batter it.

The blinded shark snatches the injured seal in its mouth, taking a sample bite. It has four rows of teeth in each jaw, so even a single bite causes significant damage. The bottom teeth pierce the mammal while the top teeth carve the meat away. Being comprised of cartilage the top jaw is flexible and can be detached from the skull, allowing it to move deep into its struggling victim. It pulls its lips back to project the teeth even further. In just one mouthful the shark can rip out twelve kilograms of blubber. The teeth are not just cutting implements but precise tools that evaluate the concentration of flesh and muscle. The tongue is even more specialised, seeking out specifically the taste of fat. Not enough and the prey is spat out again. Combined with nostrils that can detect 1 part of blood in 10 billion parts of water and you have one of the most advanced tasting mechanisms in the animal, or even mechanical, world. If the resulting taste is sufficiently pleasant the great white begins dining.

Great white shark, South Africa- ***Alberto Carrera***

Thankfully it is blubber which is most appealing to them and this occurs in higher concentrations in marine mammals than it does in people. Only once its struggling victim ceases struggling does the great white roll its eyes back into position and finally gaze at its meal.

The term 'man eater' in fact barely suits these hunters, the truth of the matter is that most people simply do not contain enough blubber to justify a kill. Indeed in the second half of the Twentieth Century, with the introduction of improved medical assistance, from the more than 160 attacks recorded worldwide from great whites barely thirty people died. Considering the average great white can weigh up to two tonnes the sheer number of survivors indicates that, even when considering improved medical aide, that people just aren't the food of choice. This is not, however, the case for most seals. Their insulating fat makes them ideal prey and several hundred elephant seals will perish each year at the mouths of the spiralling sharks.

The great white may appear formidable but it is dwarfed by fossils of an even larger species. Like most sharks the soft skeletons of these ancient giants had a minimal chance of surviving the process of fossilization but its teeth are sturdy enough to have survived the passage of several million years. The largest great white tooth is some seven centimetres long; the fossil teeth were more than eleven centimetres, more than capable of inflicting serious damage with a single bite.

The owner of these efficient weapons was fittingly called megalodon, giant tooth. It was thought for some time that this monstrous fish was the direct ancestor of the great white and thus that the great white had been dominating the seas, in a way, for quite some time. But genetic analysis has shown that the closest relative of the great white is not the megalodon but the mako, the swift shark of the open ocean. While the mako will tackle the likes of tuna and swordfish it is certainly no match for the likes of elephant seals and juvenile whales.

DNA shows that the mako and the great white last shared a common ancestor about seven million years ago. This was about the same time when the transition of forests into savannah on land forced a major split between two relatively similar species of ape. In other words it has taken the same amount of time to transform a vegetarian similar to a chimpanzee into a book reading human as it has to modify a sprinting lithe shark into a powerful mammal killer. Great whites are not the primitive monsters they are so routinely and inaccurately portrayed to be; their appearance as top predators is as recent and sophisticated as our own ancestry.

More recent genetic studies show another, surprising, shared feature with humanity. During the peak of the Ice Age, about 10,000 years ago, environmental pressures and increasingly dire climatic conditions had driven many great white shark populations to extinction, until only a single group was able to hold on, just, in the southern Indo-Pacific Ocean. It would take until the end of the Ice Age and increasingly favourable temperature conditions for the great whites to be able to dominate the ocean once more and take advantage of plentiful seal prey in South Africa, the North Pacific, and right across the Atlantic. Humans experienced our own genetic bottleneck, in this case about 900,000 years ago when little more than a thousand people were left in one tiny corner of Africa. For both species the previous vulnerability is still present in comparative lack of variety in our genes compared to other species- making us more vulnerable to inbreeding and other environmental pressures such as disease or competition from other predators - though people are fortunate enough to have had more time for our numbers to recover compared to the great white. That vulnerability doesn't just impact the sharks. In South Africa, where hunting by orcas in addition to overfishing have reduced the previous apex predator to nearly redundant levels, the entire area is feeling the impact. Seal numbers have exploded- placing pressure on local fisheries that is unsustainable long term, while other sharks have started increasing their patrols- species that unlike the great white are more than happy to not just take a mistaken bite out of a surfboard but are happy to snack on its occupant also. A world without the great white might not be the nirvana it initially promises.

Great whites are not the only fearsome looking hunters in their group. Stalking the coastal waters of Africa, Australia and the Americas is an unusual snaggle toothed shark, nearly as large as the great white. The giant has several common names including the grey nurse or sand tiger shark even though, confusingly, it is not directly related to either tiger sharks or nurse sharks. There was equal confusion in the wider public about the grey nurses' diet resulting in it being misidentified as a man-eater and hunted until the edge of extinction, even though a single check of its teeth would have indicated it could not tackle successfully anything much larger than a salmon. Like its notorious cousin the grey nurse is not the blood crazed menace it at first appears. They have an unusually tender characteristic; like many sharks they give birth to live young, nurturing them inside their body until they are developed enough to survive on the outside. But the arrangement is not as family friendly as it may at first appear.

Grey Nurse shark, South Africa- ***Alex Vog***

Most fish that invest in live birth either have extra-large yolk connected to their young or continue to supply nutrients to their growing infants in a way similar to the placenta used by most mammals. The grey nurse uses neither strategy. Instead, when the developing sharks begin to hatch inside their mother the only food left for the toothless young is the last pitiful remnant of yolk in its own egg and that is quickly consumed. But their mother's body does not release them; they are trapped inside her for now. Nor are there just a few famished infants inside their mother. Over the last couple of months the mother's body has stored multiple egg filled capsules inside her. But not all eggs hatch at the same time. The hatchlings might be only newborns but they are still sharks and the first to hatch feeds on the unhatched eggs inside its capsule. Even this only feeds the young for a few days and soon the starving sharks have no option left but to break out of their capsules and seek more food.

Finally freed from their capsules there are now close to fifty infants squeezing around inside their mother but even now she doesn't deliver them and the eggs she keeps producing are not sufficient to feed them all. By this stage the young are each about ten centimetres long and already growing their first sets of teeth. Having no other option the juveniles feed on each other. She will not go into labour until her young are at least a metre long. Only two infants will make it out of their mother alive and even this number is only due to the grey nurse having two uteri like all sharks. Another relative, the crocodile shark, puts some restriction on the level of fatalistic conflict between its offspring by delivering when there are still two young left in each uterus.

Not all sharks are such fierce hunters from birth or at any time of their life. The megamouth, for instance, only has four teeth in each row and these would have trouble cutting through lettuce let alone flesh. It does have about a hundred rows of teeth though and these are effective for snaring smaller prey. Its size cautions you to keep your distance for the owner of these peculiar teeth can reach five metres, larger than some great whites. It is still barely understood, despite its girth it managed to remain hidden until 1976. Its most notable feature is its cavernous jaws that expand like balloons until they are wider than the shark's body and then collapse with a healthy portion of plankton caught inside. These peculiar jaws have something else about them which catch your eye. The megamouth, like the cookie cutter, attracts prey with a light organ.

Whale shark, Contoy Island, Quintana Roo, Mexico-*Imagesourcecurated*

The light organ of this mysterious giant is located not on its skin but inside its mouth. In the darkness where the megamouth normally hunts it is possible that the luminescence disorientates preys and brings it straight into the shark's open jaws.

Other filter feeding mackerel sharks grows even larger than the peculiar megamouth. The basking shark, with its preference for cool plankton rich waters we have already encountered, but in warmer waters it is outcompeted and outclassed by yet another giant, the whale shark.

Plankton is not the typical meal for a shark and the effects of this change in feeding patterns means that the whale shark's closest relatives are not immediately recognisable. It is certainly no mackerel shark; instead it is a giant relative of wobbegongs and other cat sharks. Regularly reaching fifteen metres it is the largest of all fish although its uppermost size is, as with many other oversized fish species, fiercely debated. According to some it can be eighteen metres, others measure it as longer than twenty metres but characteristically the biggest ones always seem to get away. Its mouth alone can be a metre long, lined with three hundred rows of teeth.

The combined total of teeth is more than that in any other kind of shark, including the great white, but these are so small that they are almost non-existent. In order to aid in sucking up small prey the mouth is placed at the very tip of its head, rather than underslung like it is in other sharks. This massive fish appears to be of little threat to anything much larger than a shrimp and so it was commonly believed for many years to be a harmless exclusive planktivore. Yet despite this the local fishermen of the Indonesian province of Papua, in New Guinea, complained that the sharks were emptying their nets of their catch. Such accusations were believed to be false, as it was considered a proven fact that these large fish ate nothing more than plankton. Such previously irrefutable facts are now being reconsidered.

In the ocean vast shoals of plankton can easily become dispelled and spread beyond reach of the whale sharks. Any isolated swarms of plankton are barely enough to maintain the hunter for more than a few weeks. It is now that the leviathans begin acting like true sharks, taking advantage of any meat they can track and pin down. Often small baitfish gather around the shark when threatened by passing tuna and typically they are safe in their choice of shelter. But with plankton this scarce the whale shark displays a new, unexpected, behaviour. It dives, leaving the helpless baitfish surrounded by a host of predators. The tuna start ripping into the vulnerable shoal but the shark is also preparing to take advantage of the assembled feast. With surprising agility for such a giant it turns around and begins torpedoing through the tightly gathered mass of fish. It plunges right through, taking tuna and baitfish with little preference, all disappearing easily into its mouth. Occasionally it may snatch up a fish that is too large for even its mighty body to process. With such inadequate teeth it is unable to chew its meal so the only solution in this case is to turn its gut inside out, emptying its contents into the ocean. Perhaps it was the sight of such feats that inspired biblical tales of giant fish swallowing up men and spitting them up whole. For the most part whale sharks are not a threat to people, except in a few special places like coastal New Guinea where fish and man compete directly.

Like their ocean dwelling relatives the whale sharks here are prone to face starvation when the plankton bloom either migrates or fails entirely. With their sensitive ears they cannot help but hear the frantic struggle of alternate prey; millions of small sardines and the like trapped by predators. But in this case they are not pinned in by tuna but by a vast net, which is rapidly leaving the water. The whale shark acts rapidly and unexpectedly. It rises, seizing the net in its jaws. Its mouth might be ineffective when compared to most of its kin but its formidable weight of twelve tonnes well and truly compensates for this. Often the net will partially tear and a few fish slip out- into the waiting mouth of the behemoth. Occasionally one of the older nets will be ripped apart by the force and the sharks will feed enthusiastically on the released bonanza. Such raids provide enough picking for several sharks and about half a dozen often reside around these popular fishing grounds year-round.

There are other places where whale sharks overcome their normally solitary behaviour. Off Ningaloo Reef in Western Australia several hundred sharks begin to gather in March. In groups it becomes clear that the fish have their own distinctive patterns, each with their own unique arrangement of spots. Due to this different sharks can easily be recognised and it is likely that the animals can distinguish each other in the same way. Their size restricts them from entering the reef itself but the particular meal they are waiting for will come to them. Over the next couple of months millions of individual coral will simultaneously spawn, releasing an immense cloud of eggs and sperm that will wash out to sea. This is what the sharks are waiting for and they plunge eagerly into the cloud, straining out a meal that will support their immense girth over the coming months. Despite their apparent energy they maintain an exceptionally relaxed physiology; with a pulse as little as twelve to sixteen beats a minute when they are completely relaxed- such as when a researcher is hard at work rubbing copepod parasites off the shark's lips; it is a little bit more difficult for people to check observations properly when the world's largest fish is in a hurry.

Whale sharks are as prolific breeders as they are feeders. Resting safely inside its mother can be up to several hundred eggs. These are some of the largest eggs known in the world, up to 35 centimetres long. These eggs are thought to hatch before the female delivers, allowing the juveniles to be a little more capable of independent living before birth; indeed one individual caught in the 1990s had three hundred pups of different ages inside her. Scientists off Galapagos noticed that all the females gathering there already had swollen bellies and came up with a novel way to examine the sharks further- giving them an ultrasound. Using a portable waterproof ultrasound was the same way that the whale shark's beating heart was able to be observed at Ningaloo but these results were if anything even more impressive.

The bulging abdomen was an accurate indicator of the female's gravid condition but the eggs that were picked up appeared to be unfertilised. Curiously the majority of the males around the islands were not yet mature so they were unlikely to be able to fertilise any of the eggs; indicating that perhaps the female stores sperm from previous mating. The only hint we have of their courtship is from a single sighting in Ningaloo Reef of a mature male biting a young female on the tail, everything else remains a mystery. For the world's largest fish we know surprisingly little about the majority of their life history.

The whale shark looks so superbly designed to a lifestyle scoffing plankton at the surface that it is hard to consider that it has other distant relatives which barely venture above the sea bed their entire life. They have no need for streamlining or prodigious appetites and are equally successful. About 300 million years ago their ancestors, a different group of sharks, split off to pursue this slower lifestyle becoming in the process fish so different from their relatives that few people would recognise them as being even remotely related to sharks; even biologists give them their own group. They are the rays.

The Australian angelshark, though very much a true shark, gives an indication of the first changes that the ancestral rays may have made to their body shape. Its body has begun to flatten out. The pectoral and pelvic fins are wide but not yet to the extremes of most rays. It also feeds on bottom dwelling invertebrates, the staple meal of many rays and part of the reason why they have developed such a ground hugging lifestyle. During its entire life it barely moves from the seafloor, even if threatened it does not swim away but instead turns and viciously bites its attacker.

Some rays, like the guitarfish, look remarkably similar to angelsharks. There is one crucial difference. The process of flattening that the ancestors of the ray took resulted in a body that was almost two dimensional. Unlike a shark the sides of a ray do not blend smoothly into each other but jut out abruptly. This has meant that a ray's gills, and often its mouth, are on the animal's underside.

Australian Angel Shark, Wilsons Promontory, Victoria, Australia- *Philip Watson*
Common Guitarfish, Spain- *Luis Pérez Berrocal*

Spiracles aid the ray in breathing while its head is buried in the soil, just as they do with epaulette sharks, although the ray has to tilt its head to one side so that the waste air can come out the gills. The lowered mouth also has the side effect of the ray being forced to eat blind. The white spotted guitarfish still looks rather uncomfortable with this arrangement. Not only is the mouth below its head but its entire head tapers to a beak shaped point further impeding its ability to see what it is eating. To compensate it has a habit of tilting its head and upper body right up so that it is capable of looking a human swimmer, quite disconcertingly, right in the eye.

A close relative, the shovelnose ray, has a body that enables it to see its food before swallowing. At the tip of its long snout are two large transparent triangle sections of skin which allow it to peer through its own body to get a clear view of the sand and anything else below. The transparent skin can also double as effective camouflage. If any sharks pass by, the shovelnose slides under the sand, leaving just its head to blend seamlessly into the background, and keep an eye on the progress of the hunter.

Another of the guitarfish's relatives, the eastern fiddler ray, also relies on skin tone to keep it safe. Rather than being transparent this ray has vivid stripes of pink and greyish brown running over its body. It may appear conspicuous when swimming but again if threatened it slips seamlessly to the seabed to hide, this time not in the sand but within the seaweed. Solid skin may act effectively in concealment but can also prevent the ray from seeing what's under its mouth. Like sharks rays have highly effective sensory equipment and use this to pinpoint a meal precisely and avoid swallowing sand instead. Some rays use this electrical sensitivity to hunt in a way peculiar to this group. They not only detect prey they electrocute it.

While many marine fish can detect the electrical vibrations of other animals, controlling the production of their own electricity is another matter. Salt water tends to impede the flow of electricity. Even the sharks with their highly electrically sensitive snouts have not developed this form of behaviour. But electric rays have a pair of specialised glands behind their head designed to produce electric waves capable of shocking passing creatures. The glands are massive and, in some species, make up a third of the animal's weight. Through sheer size the rays are able to overwhelm the deadening effects of the water and send a zap into an approaching predator. Even a shark will change course if it encounters a large electric ray during its nightly hunts. Firing electricity into the body of an approaching fish is useful not just for defence but for attack. The Atlantic torpedo produces the strongest shocks at 200 volts. It doesn't rely on a chance shock to kill either. It wraps itself around its target so that its shocking is delivered directly to it. At close range such a shock can stop an animal's heart.

Other electric rays aren't large enough to hug and shock their food to death. Numbfish off the South coast of Australia lie on the seabed waiting for suitable prey to pass. Staying on the one spot for several hours encourages small invertebrates to investigate. One of the most peculiar is a worm that lodges itself in the ray's body, typically in the spiracle behind the ray's eye but occasionally it hooks onto the eye. This doesn't seem to bother the numbfish as it doesn't use sight much for hunting. The wriggling worm attracts other attention and small fish are easily drawn towards it. As they hone in on what looks like an easy meal they ignore the jaws opening beneath them. In an instant the fish has been zapped, stunned and snapped by the numbfish.

Electric rays contain the smallest of all rays, the short-nose at a tiny 10 centimetres across. It feeds on small invertebrates and whatever else it can catch. Unlike sharks no rays have oil filled livers to maintain buoyancy; if they stop swimming they sink to the ground. Electric rays rely on their powerful tails to provide the momentum to drive them forward, in a similar way to a shark. Guitar fish and fiddler rays swim in the same way. But most other rays have such narrow tail fins that they are more of an impediment to swimming than aid. At best the tail acts as an additional form of steering. A solid body provides its own way to move forward though.

The pectoral fins expand from the ray's body much the same way as the wings of a bird. These fins are supported by a cartilage formed fan-like structure which allows for firm but flexible muscular movement. By flapping these fins the ray can move with a technique that looks more like flying than swimming. Movement runs down the large fins, reminiscent of the smooth wriggling of a snake, and the continual rising and falling motion allows the ray to keep its position in mid water. With the exception of the electric rays and guitarfish virtually all rays use this Rajiform swimming style rather than rely on their tails. Consequently, when a fever of rays glide past you they give more an impression of a flock of birds than they do a shoal of fish.

Rays, for the most part, are at their most comfortable not when swimming but when hiding submerged in the substrate; some just rest on top while others bury all but their eyes underneath. Some have developed an unusual feeding technique as a result of their breathing adaptations. The constant stream of water coming through its mouth and spiracles has to leave by the gills on its underside but by tilting its body and pressing it into the sand the ray can control which side it comes out of. In this way a ray can flush out hiding prey simply by breathing. But one ray employs a rather more active hunting style, the sawfish.

While the great-tooth sawfish spends most of its life lying on the sea floor it can also be seen swimming actively through the water. When swimming it could almost be mistaken for a shark, were it not for its saw. Indeed it sometimes shares the coasts with sawsharks, sharks that share the same lifestyle and general appearance. They can be distinguished at a glance. Sawsharks have tentacles stretching along their snout allowing them to feel for food. It also has gills directly behind its head. The sawfish's snout in comparison is unburdened by tentacles or other protuberances and the gills are on its underside like all rays. Despite its shape the sawfish still swims with the distinctive ray style, relying on its flapping pectoral fins rather than its tail. Its body may have changed significantly from the basic design of most rays but its history and family relationships can still be seen clearly in the mode of swimming it has inherited.

There are seven species of sawfish, most of them about two metres long or so. But even in the largest species it is the snout that takes your full attention rather than the fish's vast body. The saw is constructed from specialised teeth that jut out of the side of the snout, as sharp and efficiently designed as the human tool of the same name. In the great-tooth sawfish, a medium sized species, there are from 14 to 22 of these teeth jutting from the snout. In the small-tooth sawfish there may be as many as 32 teeth. Only the teeth are small in this species; the small-tooth sawfish can reach seven metres long.

The purpose of the saw was a mystery until only recently. Due almost entirely to its resemblance to human tools the saw was viewed exclusively as a weapon, perhaps one to carve flesh off its victim, despite the fish rarely being observed feeding let alone hunting. Modern examining techniques were applied to the snout and came to a separate conclusion. Scientists found that the saw was highly sensitive to electrical stimulation and its odd shape helped the fish to find secretive prey, working in a similar way to the distorted head of the hammerhead shark. Sawfish in aquariums were tested to see if they could detect electric amplifiers dangling in their exhibit. They did so expertly, quickly demonstrating the effectiveness of their saw at detecting electrical vibrations of prey, which became even more obvious when their skills were compared to guitarfish which simply blundered into the amplifiers. On this basis it was considered that the saw was used entirely in the detection of prey, not hunting it- as the act of wielding it against prey would likely damage and interfere with its sensitivity. Observations of the sawfish in action would soon challenge this theory.

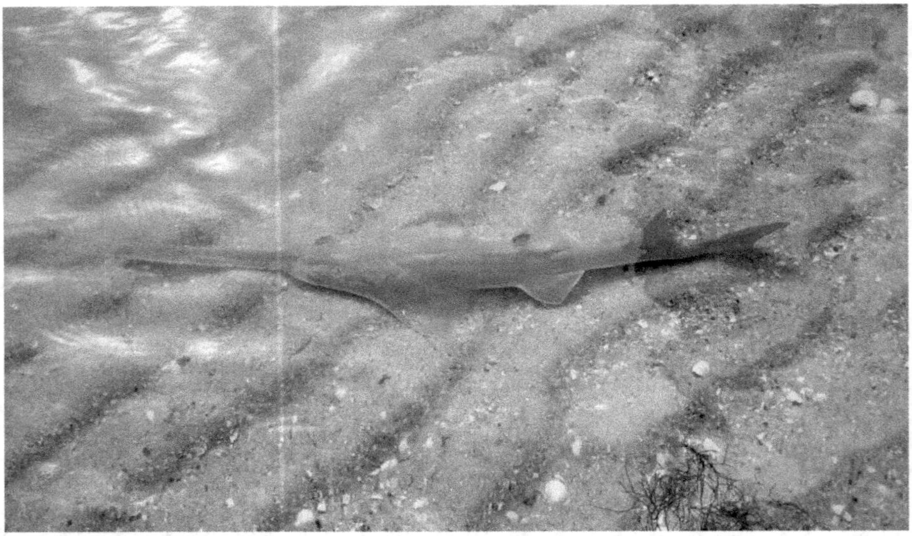

Small-tooth Sawfish, Florida- ***Patrycja Lawryniuk***

When a small-tooth sawfish detects a hiding crab it begins to shuffle its snout into the sand, unearthing the crustacean. Before the crab has a chance to flee it is pinned down by the tip of the saw. The fish begins to slide itself forward until its mouth is directly over the struggling animal and it can begin to feed. Swimming prey like a wrasse elicits different behaviour altogether. The sawfish leaves the seafloor and begins to trail the smaller fish. It soon catches up and begins to circle it, cutting off any chance of retreat. With the wrasse confused and unable to know which way to turn the sawfish begins to swing its saw around frenetically. The saw is not used to slice the wrasse into edible pieces as once thought but as the fish is thrashed around the saw certainly cuts into it. But the saw is a hindrance when it comes to swallowing and the sawfish has to keep battering the wrasse until its mouth is close enough to snatch it. The sawfish are the only animals alive that use the same part of its anatomy for both sensory purposes and attack, in one of the most dramatic hunting techniques imaginable.

With their oversized head sawfish are able to fiercely protect themselves from most predators. Any shark that swims too close will soon leave with a deep cut running along its flank. But sawfish have one danger that their snout cannot protect them from, if anything it is their prodigious snout that endangers them. Sawfish give birth to live young, young which like their parents are armed with sharp protuberances running along the side of their head. The unrelated sawsharks also give birth to live young but in their case the young come with their snout conveniently folded back facing away from the mother. But the sawfish baby's snout is erect from birth; it is also sharp and fully capable of inflicting injury. The mother sawfish is at risk of having her soft insides ripped apart by her own young. But like the medieval knight that stored his sword inside its sheath to avoid stabbing himself the sawfish comes into the world complete with its own sheath. Fitting loosely over their snout like a mucous made pillowcase the sheath in this case acts as protection not for the juvenile but for the mother both during pregnancy and labour. After birth the sheath slips away and the infant sawfish is suitably armed for the dangerous life ahead of it. It has no bond with its mother and disappears alone; heading for quieter and less well patrolled territories. Its parent in the meantime, continues releasing the rest of her brood. Giving birth is dangerous for the sawfish not just because of the risk of labour but because of the inadvertent release of body fluids in the process. Blood in particular soon attracts sharks. For the small-tooth sawfish giving birth to such well-developed infants inadvertently puts the female at substantial risk. Live birth is an unavoidable hazard not just for sawfish but for nearly all rays, from guitarfish to torpedo rays, stingarees to manta rays.

Only one family of rays lay eggs; the skates. Eggs are easier to produce in comparison to live young so the skates have much more offspring than other rays. With a larger amount of young being born there is a higher chance of skates being born that are a little larger, or have slightly different marking, or have wider bodies. But being born so immature also places the developing rays at higher risk of falling foul to predators. A slight improvement in camouflage might be all it takes for one skate to survive while its sibling is eaten. If it survives to adulthood some of the skate's own offspring may develop such a colour change even further and if they outlive their siblings they will again pass on their genetic advantage. Over millions of years the results of such competition between individuals is spectacular. It is not surprising that skates are the most numerous group of rays with more than 200 species.

Young thornback skates feed and shelter mainly in the shallows. For an adult there is larger prey like crabs and flatfish further out to sea but this makes the ray an obvious target for sharks in turn. Thornbacks have a line of short spines running down their spine as defence. The tail is an even more elaborate weapon. It stretches back like a whip, covered with barbs that become increasingly larger near the tip. The tail can be more than twice the length of the ray's body. Its tip is armed with at least one and occasionally two venom laced barbs. The thornback has taken the first step towards one of the most effective of all defensive strategies. Several other families of rays have also seen the advantage of arming their tail with such efficient weapons. The stingarees are some of the most common. They have round bodies like saucers, only saucers with a diameter of eighty centimetres. The sting is effective but for the most part stingarees rely on their awkward shape to avoid being swallowed. There is another family that stingarees are often confused with which has taken the development of barbs much further, the stingrays.

Stingrays' backs are smooth and thorn free, allowing them to glide easily through the water. The tail however may be tipped with a single brutal barb or seven. A quick swish propels the barbed tail easily into the attacker's flesh, ripping the barb out in the process. The barb is designed like the blade of a steak knife; ideal for driving forward but if an injured victim tries to pull it out the points of the blade snag against the flesh. Simultaneously the barb is also leeching venom into the wound, further incapacitating the threat. The stingray readily recovers from its loss, growing a replacement barb at the rate of a few centimetres or so a week. Indeed if a stingray doesn't use an old barb it will eventually be pushed out by the new barb developing beneath it.

Stingrays, unlike the sharks, have conquered not just the seas but fresh water. South America has become the epicentre of freshwater stingrays. Every couple of years or so a new species is found and many of these are large. They also use vivid patterns to stand out.

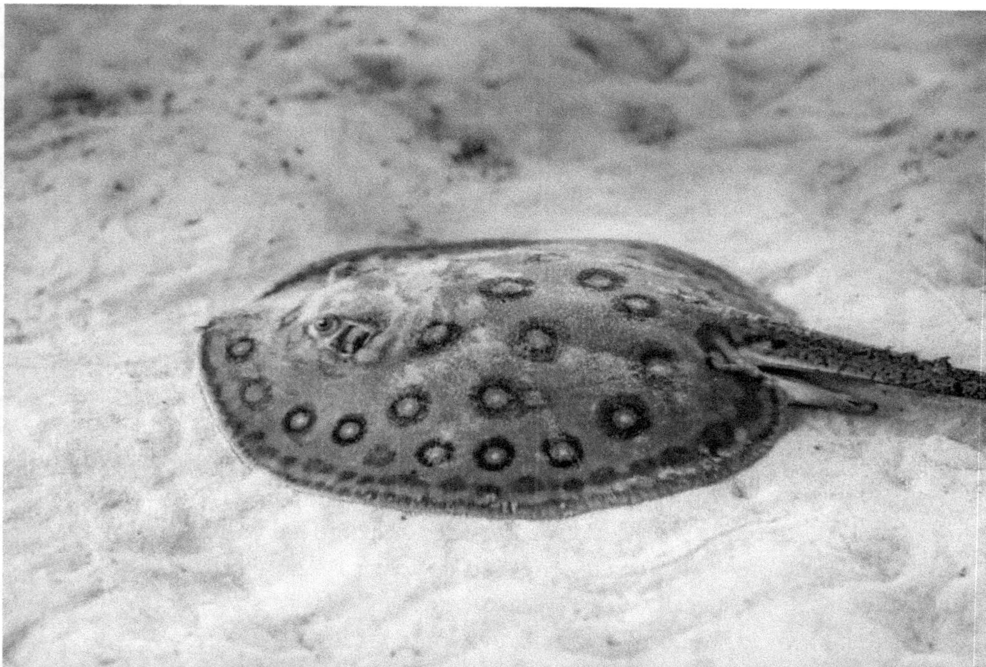

Oscillated stingray- ***Diegograndi***
Blue Spotted Stingray- ***ImageSourceCur***

In Venezuela the local species of ray has a mosaic of brown cylinders separated by a mauve background running across its back. In Paraguay the oscillated stingray has red spots, each with a strong black outline, splattered over its body. The stingrays need to stand out as in these waters caiman and even jaguar will attempt to catch whatever fish they are lucky enough to encounter. Only by clearly advertising that they are dangerous, does the ray reduce its risk of being attacked. It is such an effective form of defence that some stingrays still living on the reef, such as the blue spotted stingray, also use it- for even though sharks lack colour vision the warning pattern itself remains visible, and other rays are able to use the vivid colours to distinguish between species.

Adult stingrays may be able to cope with the challenges of river life but the same is not the case for their young. Even though all stingrays give birth to live young these infants are typically not developed enough to cope with the mechanical and anatomical problems of fresh water, in the same way as fish like piranha and guppies that ancestrally lived in fresh water. So, freshwater stingrays devote a little extra energy to their developing young.

As their offspring wriggle inside them ready to hatch the mother produces uterine milk. This milk nourishes the young in the same way that the placenta does in mammals. With this extra surge in nutrition the young rays are as prepared as they possibly can be for the outside world. They are now free to grow to prodigious sizes. The largest of all these freshwater rays, the Mekong stingray, lives in Cambodia where it may reach four metres across its outstretched fins and is one of the largest of all freshwater fish. Being such a size means it has little need to advertise its potentially dangerous nature and instead the back of the ray is a dark muddy brown colour.

But living in a river, even one as large as the Mekong, restricts the growth of a fish. This is not the case in the ocean however and some have left the protection of the coastline and grown proportionally. The pelagic stingray is the largest of the stingrays, about a metre and a half, and it needs to be. As it travels across the ocean it may be forced into fasting until it encounters sporadic large congregations of food, like a passing shoal of baitfish. Occasionally it may find a real bonanza like squid spawning just off shore. The stingray takes advantage of its large fins, using them to batter prey into its waiting mouth. In the peak of a hunt it will even turn upside down if it needs to; allowing its upside down mouth to finally be able to reach any fish sheltering at the surface.

The pelagic stingray frequents waters so deep that it is rarely able to rest on the bottom like other rays. Should it come across a shark or sperm whale it can defend itself with its spine but it has a higher chance of survival if it can outswim them, or at least outswim any of its companions. But having a long spine dangling from its tail causes significant drag, most noticeable when the ray needs to escape from danger. So some relatives of stingrays have streamlined the design by losing their sting altogether.

The cownose ray rummages through the sand near the coast; its almost completely smooth tail flicking proudly behind as it snuffles up any small prey it can find. The cownose still has a sting but it is feeble when compared to its stingray cousins. The snout of this fish hooks up peculiarly so it can better suck up crabs and other burrowers, giving it quite a resemblance to its grazing terrestrial namesake. It is not as big or dominant a predator as the stingrays so it can afford to live in groups. A dozen or so of these brown rays churn up the ground beside each other. This reduces the amount of food available but the cownose rays have more immediate concerns than starving. With small stings and equally small size they would make an easy meal for a shark.

Feeding in larger numbers reduces the risk of any individual cownose ray being singled out and attacked but gregariousness does not come naturally to them. This shows up clearly in their positioning; they are not as attuned to each other's movements as natural schoolers like sardines or anchovies nor do they naturally move into a synchronised formation when threatened. Nevertheless, when conditions are right, they can gather into shoals of more than five hundred, with tails flicking through the water with such a flurry that they can trigger mistaken shark alerts on land. On a few rare occasions these groups can reach a thousand rays but such a sight quickly disentangles into the deep as the rays split up once more to feed.

Cownose rays have larger relatives combing the seabed. Spotted eagle rays are among the most spectacular of these. When measured from the tips of the pectoral fins, or wings, they can be three metres across. As they rise gracefully to the surface their white spots contrast easily against their aqua green bodies. Their mouths are more rounded than the cownose rays indicating a different style of feeding. It rests against the shore and begins to pump water out its mouth into the sand. With each puff it blows away more sand, quickly uncovering any hiding sheltering worms or cuttlefish.

◄Shoal of Spotted Eagle Rays, Cancun, Mexico- *Imagesourcecurated*
◄Manta rays filterfeeding, Maldives- *Haveseen*

With unexpected speed it snatches up the discovery and begins the task of breaking the animal's body down with its seven rows of teeth. These teeth are small but tough, especially suited to dealing with the hard armour of unearthed crabs. There is indeed a mass of fish and crustaceans sheltering close to the shore but these require either speed or specialised techniques like that of the eagle ray to capture. One family of rays has taken to a simpler life; sieving minute life with massive jaws. So yet another group of cartilaginous fish have joined the basking and whale sharks as giant plankton feeders.

Large plankton feeders have a significant advantage over their competition and indeed these planktivores are the largest of all rays. To make feeding more efficient their mouths are not hidden beneath the head, like nearly every ray, but instead are positioned at the front of the head, much like a shark. Most of this family funnel plankton into the mouth with fleshly lobes that pop out either side of their jaws. When these lobes are rolled up they look like the horns on medieval demons, consequently the whole family was named devil rays despite their harmless diet and attitude.

Some of the devils rays have grown nearly as massive as the whale shark. The largest of the group are manta rays. Their delicate belly is white but their back is a majestic black that seems to cover the giant like a cloak. Their tail is fine and delicate, lacking a poisonous spine and playing no part in steering. Consequently the manta's slender tail is easily damaged and the length of the animal is not a very accurate indicator of size. Most rays are therefore measured from their large pectoral fins. Manta rays can stretch more than nine metres from fin tip to fin tip.

Being so large the ray soon attracts parasites that cling to its soft body. The manta ray deals with its lodgers more dramatically than a numbfish does. The fish uses its powerful fins to power itself through the water. Building speed it launches itself into the air. In one of the most dramatic sights in nature nearly three tonnes of fish become momentarily weightless before the manta crashes back into the sea. The force of the ray's splash is enough to remove most pests from its body. So powerful is the force of impact that some mantas have been observed leaping during labour and delivering young during the process.

Mantas can also prove athletic when it comes to feeding. Plankton may be small but it's not motionless. Each individual zooid and other microscopic animals slowly whip their fins and tiny legs through the water in an effort to escape. The enormous rays plunge through the water, taking great somersaults directly into their prey. As they rotate the sheer size of their bodies gathers all the plankton together. As they come out of the turn they can swallow the food in one easy mouthful. In good times several manta may gather together to feed and adapt another technique. They start below their target then begin circling it; each ray following behind the other, surging through the waters with such power that they create a whirlpool effect or, as scientists have taken to calling it, a cyclone. As they rise they draw their bodies closer, using the power generated by their bodies to move the plankton into increasingly tighter concentrations- until they swoop in to feed.

The manta rays have another, less energetic, feeding technique which is if anything even more stunning. They trail each other gracefully, like geese travelling in V formation to improve efficiency. Each ray is using its acute electrical and physical awareness to position itself perfectly around the rays in front of it; just as a well organised choir will have everyone positioned so that all are visible, and all have the same portion of space for movement when they need to. The efficiency the rays wish to improve is not just movement but also dietary based. As the lead ray takes its fill more plankton spills out behind only to be gathered by its companions. The fish continues feeding peacefully, barely noticing as your insignificant shape slips beneath it. As the manta passes over your head it momentarily blackens out all light above it. Then it continues on its relentless journey into the deep as you gasp at the surface trying to regain your breath. Such a magnificent fish may continue gently harvesting plankton for twenty years. Sharks and rays are truly, not just some of the ocean's most efficient predators, but their most awe-inspiring.

Chapter 3. The First Bones

Spotted Chimaera, Three Tree Point, South of Seattle, Washington, USA- ***Sara Thiebaud***
Longnose Chimaera- ***NOAA Okeanos Explorer Program-Gulf of Mexico 2012 Expedition***

Sharks and rays have a group of relatives that are poorly categorised and hence rarely spoken about. These are neither swift hunters nor gentle plankton-eaters circling gracefully below the water's surface. For the most part they are hidden in the depths seeking any prey slow enough to be snatched up by them. Their heads are boxlike, placed roughly on a muscular body, not unlike that of a salmon. Technically the group are called chimaeras after the monster in Greek mythology made out of several different animals. The name is even more suitable once you consider the chimaera's skeleton.

The chimaeras' fiercer shark relatives have a body constructed entirely around a framework of cartilage or gristle. The other major group of fish, virtually everything that isn't a shark or ray, by comparison have a skeleton reinforced with calcium, causing it to form into solid bone. Chimaera anatomy falls roughly between the two; the majority of their bones are comprised of calcium but the remaining cartilage portion is substantial enough that they can't truly be considered as bony fish. Their upper jaws, on the other hand, are primarily gristle but are fused to the skull so that they can't be projected in the same fashion as a shark. As a compromise scientists have placed chimaeras in their own special subclass of cartilaginous fish, still vaguely part of the sharks and rays group but also recognisably as different from its relatives as a kangaroo is from an antelope. If any animal can shed light on the nature of the shared ancestor that must have originally existed between bony and cartilaginous fish then it must be this one.

The group has abetted as much confusion for the general public as they have for scientists. A glance at the common names for chimaeras reveals this, titles as varied as rabbitfish, ghostsharks and ratfish. The greatest visual feature of the fish is the end of its head and chimaeras are roughly divided into three groups accordingly. Short nosed chimaeras, such as the spotted chimaera, have a blunt ending to their snout with a mouth underslung like a shark. Also like a shark the spotted chimaera has sensitive pores running around its mouth to detect the electrical signals of passing prey but. The Spookfish looks even more peculiar. This second chimaera group is notable for the protrusion jutting out in front of its head like the horn on a unicorn. This organ appears quite an encumbrance, particularly when it comes to swimming. Awkward as it might be this organ is also exceptionally sensitive to electrical signals and acts like a large sensor, drawing the spookfish's attention to the presence of other fish- including predators. Together there are about thirty species from the short nose and spookfish families spread inconspicuously throughout the world's ocean. The final group of chimaeras is smaller but because of the behaviour and extreme appearance of one of its members, the ploughnose, this family is the best known.

The ploughnose chimaera spends most of the year in the depths of the ocean around Australia. You can recognise it instantly from the trunk like protrusion sticking out over its mouth which it inspired its other common name, the elephant fish. Hidden just behind the trunk two massive fused teeth sticking out of its mouth like a peculiar caricature of Bugs Bunny. But this chimaera looks strange even before you come close enough to make out the details of its face. It gleams like polished silverware, without evidence of the slightest scar or scale. Infant ploughnose have a body covered in rough denticles just like a shark but as they mature the denticles slowly break away until the fish is left with perfectly smooth, scale less skin. To cap it all off this species has vivid green eyes which reflect brightly under any artificial light.

The ploughnose is an interesting enough fish patrolling the darkness but once a year they come close enough to be watched not through the window of a submersible but by the torch of a diver. The chimaeras travel to the coastline of Eastern Australia in their thousands. Simultaneously other populations are also heading to the coasts of Africa and New Zealand. There is little suitable for adults to eat here but the nutrients and microscopic invertebrates will cater ideally for their offspring.

Like sharks they fertilize their young internally but with their unusual jaws the males are unable to maintain an oral hold of the female during mating. Instead they release an additional clasper from a hidden groove in their forehead that clings to the female like Velcro. This club like structure reveals another, completely unexpected feature, it is coated with shark-like teeth and research has confirmed they are genuine teeth, a finding that has drastically changed our view of backboned animals; who would have thought that teeth would be capable of growing outside of the mouth? Once the male chimaera has the female in tow he is able to manoeuvre himself into position and get on with business. A few weeks later the females release a few large flask shaped eggs before returning to deeper ocean.

Ploughnose chimaera, Mornington Peninsula, Victoria-*Mandy Gardner*

Ploughnose may migrate to the shallows to breed but they can only go so far into the sunlit waters. The sweet fresh waters that flow into the sea are taboo to them also. Cartilaginous fish are the supreme hunters of the ocean but here they stay. After more than four hundred million years sharks and rays remain virtually marooned in this single habitat. Even today only a few species of freshwater stingray, a few river sharks and the bullshark have ventured inland. The colonisation of fresh water has been left on the whole to one group; ray finned fish. These are bony fish but they have other adaptations as well as their bone structure to help them adapt to these conditions. Unlike sharks the ray finned fish generally have a cover over their gills. This not only protects their vulnerable breathing apparatus but aids their streamlining. Their most advanced feature however is their fins. They fan out from the body supported by thin fibrous struts- unlike the solid, but comparatively inflexible fins of sharks, chimaeras, and rays. This makes them extremely manoeuvrable allowing them to do things that cartilaginous fish can't.

One of the ray-finned fishes' specialties is reversing by flicking their fins back with enough force to change their direction. A quick extension of the fins is enough to stop the fish in its tracks, whereas a shark would have to swerve to avoid crashing into an obstacle. Perhaps the most sophisticated swimming skill of ray finned fish is visually one of the simplest, hovering. The fundamental difficulty of hovering is shared by both animal and machine; the objective to stay in the same place but if either stops moving they will instantly plummet. The fish has to maintain just enough momentum to keep moving but not enough that its position changes. The trick to this comes down once again to those sophisticated fins. Through repeating a gentle semicircular motion during which the fins are alternatively brought forward then back the fish is able to achieve this remarkable feat.

Ambidextrous fins are a definitive feature of ray finned fish. So, when did these remarkable creatures first appear? The first remains of these fish date back to the middle of the Permian, 280 million years ago, when sail backed reptiles like Dimetrodon dominated the land well before the dinosaurs. The Permian was in fact a time of great change with new groups developing across the world both above and below the water. This was until 245 million years ago when the continents on their relentlessly sluggish movement over the surface of the earth slammed into each other. The result was a planet comprising a single continent, Pangaea, an impenetrable barrier to climate and life.

With all the land joined together the life giving moisture of the ocean was restricted to the coast. Under these conditions any inland tropical forests began to dry out and fail. The exposed rivers quickly evaporated. Across the mass continent once lush land was transformed into desert. Trees do more than just produce oxygen. They also harbour bacteria. When blown by winds into the atmosphere the bacterium forms a nucleus for raindrops to form around. Without the essential bacteria heavy rainclouds would drift over the skies without breaking. This lack of rain created a great fluctuation in climate and weather conditions across the globe. There is still debate about whether the conditions led the earth into either a global ice age or heat wave. One fact cannot be debated however, the sheer rate of extinction. On land this was the first and only extinction event to serious threaten the insects. Eighty three percent of their species was lost. The devastation in the oceans was even more severe as complete ecosystems like sea lily fields disappeared completely. At least 95% of all life was lost during the Permian mass extinction. By comparison the mass extinction that claimed the dinosaurs, pterosaurs and ammonites and allowed the rise of the mammals wiped out about half of the planet's living creatures. The Permian created the most severe conditions for life since it began.

Nowhere would the pressure on life be more extreme than in freshwater. With no chance of redeeming rainfall the dwindling pools would either evaporate completely or become so salinized that some lakes would have become saltier than the ocean. The closest existing example of these conditions can be found in the Arabian Peninsula. The waters of the Dead Sea have salt at such high concentrations that people are able to float effortlessly on top. This is the only significant body of water that has never been colonised by fish. Yet it was in waters like these that the first ray finned fish would have survived the Permian extinction event. This may sound extreme but all of life today, including people, can trace their heritage back to the 5% of survivors from the Permian extinction event, if their ancestors had not they wouldn't be here. Yet people and most other animals have changed radically since that time. There are, however, ray finned fish alive today that are nearly identical to their resilient ancestors.

Rooting around in protected rivers of the northern hemisphere is a most unusual fish. A pair of amber eyes glistens above a snout that sticks out like the extended nose of Pinocchio. The body is armoured like a tank and proportioned like a stretch limo. It is a sturgeon, one of the most distant of all living bony fish. Like a chimaera the sturgeon has a skeleton composed of cartilage and solid bones. Unlike any other bony fish sturgeons have five rows of bony plates running over their spine. The plates act in a similar way to the denticles on a shark, providing in-built armour.

Sturgeons appear in a variety of sizes, although most conform to the same basic shape. The European sterlet is one of the smallest and most recognisable of the group. They are named for the shape of their bony plates which vaguely resemble a childlike scribble of a star. The rest of the sterlet's body is pretty standard for a sturgeon. Its mouth is placed below the snout so, like a ray, the starlet is unable to see what it's eating. Rather than sensory facial pores sturgeons have four barbells, taste sensitive extensions, wriggling beneath the mouth. Molluscs sheltering between stones make a particular treat. Once the barbells have detected a potential meal the starlet unearths it and begins to feed.

In North America a relative of the sterlet, the shovelnosed sturgeon, has an even odder shaped head. It has developed a blunt snout that if anything resembles an ironing-board more than a piece of digging equipment. It may look like an encumbrance but its unusual shape creates a larger surface area to detect food with. Compared to the rest of the head the mouth is surprisingly primitive. It is in fact little different to that of a lamprey but unlike its distant relatives no sturgeon is a scavenger. Anything that has not been freshly killed will be discarded. They prefer to snatch up shrimps, small fish and freshwater snails.

Sturgeons are picky about not just what they eat but when. Winter in the northern regions of their range brings a dramatic drop in temperature. A sheet of ice soon covers the top of most lakes and rivers. In larger bodies of water the ice only stretches down so far then stops, where it begins to act as an insulator for the remaining liquid. The bottom of these lakes may not be cold enough to kill a fish but it still saps their energy considerably. To survive shovelnoses begin to feed selectively on other small fish sheltering with them, using the fat in its prey's body to fuel their own. There is another reason that sturgeons have to be selective feeders. They don't have a stomach. Instead, like the sharks, rays and chimaeras, they rely on a spiralling valve in their gut to obtain all the nutrients in the food they need.

The shovelnosed sturgeon shares another feature with the elephant nose chimaera; it migrates to the same area annually to breed.

Sterlet- ***Rezkrr***
Sturgeon- ***Estproject***

Each sturgeon species has its own preferences when it comes to selecting a brooding site. Shovelnose tend to deposit their eggs in the gravel of swift, flowing streams. Here the water is well oxygenated, carrying insect larvae from upstream with the current, ideal food for developing young. Shovelnose larvae are already quite advanced before they hatch due to the high quality of yolk inside the egg. Such a source of nutrition also proved attractive to people, not just as a meal but for its commercial value, to be shipped off and sold around the world as caviar. Once shed the eggs are difficult to collect and relatively worthless so there was only one way to collect them; direct from their source. Female sturgeons would be killed and sliced open by fishermen in order to claim their prize.

The shovelnose is relatively small so it survived the sturgeon slaughter that was so prevalent in the late nineteenth and early twentieth century virtually unscathed. Instead focus was directed on a much larger fish, the six metre long white sturgeon. Such titans could reach six hundred and eighty kilograms and were the largest fish in North America, indeed the longest of all sturgeons. Such a size does not occur overnight; indeed some of the animals were more than a hundred years old. Such large creatures were virtually an unmissable target for those seeking the liquid gold of caviar. Consequently the population dwindled significantly, an impact that is still being felt today, and indeed many current species of American sturgeon remain endangered. Other species have only recovered due to extensive conservation action. Such severe hunting was not restricted to the Americas.

The beluga once lived in rivers across much of Eastern Europe. Like its American relative it took a long time to reach maturity. Beluga can reproduce at their youngest at about thirty years of age and even then they are still relatively juvenile. This four metre sturgeon takes great care in choosing its spawning site. The females gather in deep holes in places like the Volga River. Unlike chimaeras the sturgeons' eggs and, for the most part, nearly all ray finned fish eggs are fertilized externally. Rather than inserting his sperm by claspers into the female the male rushes over to the spot where the eggs are being scattered and releases milt in a great cloud, leaving it to chance and the currents to fertilise the young. The odds of any individual egg coming in contact with the milt before it is washed away is almost zero. To make up for this the female produces a continual stream of eggs during mating. A single beluga female can contain 180 kilograms of almost microscopic eggs inside her. This rich source of caviar again came to commercial attention. Again the species was brought to the brink of extinction. This time a radical, but still invasive, technique came to the fore which allowed the eggs to be extracted without killing the female. Providing that a female is only harvested every few years the sturgeons can breed normally in the intervening years and the population has remained stable. With almost all of the twenty one known different kinds of sturgeon at risk of extinction this new technique may prove the group's final and most important reprieve.

The Russian beluga is notable for another reason other than its caviar. It is enormous. The white sturgeon may be the longest of its family but it is easily outweighed by the beluga. This monster used to reach four hundred kilograms. Part of the reason for this may be its unique mouth among sturgeons. Unlike most sturgeons its jaws are not beneath the snout. Instead the beluga's head terminates with its mouth, allowing it to scoff up food directly in front of it. This facial arrangement means that the beluga is not restricted to sniffing out invertebrates in the mud but can actively pursue and hunt fish at eye level. Having the mouth opening at the front of the skull has developed from being an innovation in the ancestral beluga to a standard feature in all other ray finned fish and, independently, the land vertebrates such as reptiles and mammals.

The Beluga has another large, although rather more distant, relative. The Yangtze is one of the major rivers in the world; running for 5,150 kilometres across China. Such a river provided ideal habitat to Giant Sturgeon for countless millennia. As they matured they would swim from their nursery grounds closer to the sea upstream, until they spawned in waters far enough away that when the eggs were washed downstream they would hatch in the same stretch of water as their parents had. These fish could reach five metres as an adult and weigh more than five hundred kilograms. But the Yangtze Sturgeon wasn't the only giant in these waters. Although most sturgeons are devoted hunters of small fish and crustaceans meat is not the only food in these waters. In the largest rivers and lakes plankton occurs in such numbers that it is easily collected by moderately sized fish. One branch of the sturgeon family, the paddlefish were attracted to these great bonanzas.

As in the oceans the plankton provided enough food to feed the largest animal in the area. And here was found one of the most remarkable of all freshwater fish, the Chinese Paddlefish or Swordbilled Sturgeon. In general design it resembled a sturgeon, except for its head. Projecting over the jaws was an extension of the skull that vaguely

Chinese Paddlefish- ***Nouvelles Archives du Muséum***

resembled a paddle. It did not use this paddle for swimming, nor for digging in the mud for prey as was once surmised.

Instead running in two rows along the paddle were curious dark pits similar to those sensory cells prominent in sharks. So similar in fact that they were given the same name, ampullae of Lorenzini, and were found to be used for the identical purpose of detecting the electrical movements of other animals. With its paddle the giant could sense the shoals of plankton in the muddied darkness. Due to its size this unique fish was heavily overfished and, like the beluga, aggressively harvested for its caviar before many specimens could be adequately examined. So the largest verified record of this fish is three metres long, still a considerable size but minimal compared to the size of the individuals that would have been caught prior to their wholesale slaughter. The highest estimate of the paddlefish's size is up to seven metres long.

A seven metre fish however takes a long time to grow and is highly vulnerable to any changes. Even though its eggs were considered a poor grade of caviar they were still collected and as the human population around the Yangtze grew by the millions so too did unsustainable pressure upon the behemoth. Like all sturgeons the paddlefish's breathing technique is primitive when compared to most other bony fish so they were particularly vulnerable to increases in pollution and chemical run off entering the river. Small dams began to divide its territory and isolate individuals, making them more visible to fishermen. The last successful spawning in the wild occurred in 1995; after that juveniles were seen no more. Yet some adult paddlefish were still seen and photographed as recently as 2003. Then, at the end of that year, their final death knoll sounded. A massive dam, the Three Gorges Dam, commenced its third stage of construction across the river, completely blocking access to the paddlefish and giant sturgeon of their breeding and feeding grounds. The giant sturgeon was caught and reared in tanks so that its high quality caviar could still be exported as it could no longer reach its natural breeding sites upriver. The paddlefish was not so lucky. In 2006 after an extensive search the Chinese paddlefish was declared functionally extinct. Not a single one had been found alive. Any that possibly survive, and not one has been recorded since that date, can no longer reach their spawning grounds and will likely leave no offspring when they are gone.

The Chinese paddlefish does have a smaller existing cousin that is still, thankfully, relatively common, the American paddlefish. Like its relative the American paddlefish lives entirely in fresh water and so would have been unable to cross the Pacific Ocean that now separates the two continents. But in geological history the location of these landmasses was vastly different. After the supermass of Pangaea that the ancestral sturgeons inhabited began to break up the continents began to form into two distinct land masses. In the Northern Hemisphere the continents that would later become Asia, Europe and North America were fused together into the super continent, Laurasia. Fossils of prehistoric ancestors of crocodiles and mammal like reptiles conclusively show that not only were these continents joined together but rivers also trailed the megacontinent. A few hundred million years ago Laurasia began to break apart and the two paddlefish species became isolated from each other. Since then there has been enough time for the dinosaurs to evolve and become extinct and for mammals to develop into animals as diverse as whales, bats and hyenas. Yet despite being separated for so long the two paddlefish have changed so little that not only can they still be placed in the same group but they are similar enough in appearance to share a common name. For whatever reason the primitive form of the ancestral paddlefish has been efficient and adaptive enough that it still works in the modern world basically unchanged.

Swimming with an American paddlefish it becomes immediately clear that this is an ancient beast. In its main haunts in the Mississippi River it has little need for speed and its movement is so stiff and inflexible that it looks almost robotic. Its body resembles that of a shark even more so than a sturgeon does, reminding us that all bony fish can trace their descent to a cartilaginous fish not dissimilar to a shark.

Yet when it comes to its senses this is one highly advanced creature. Unlike its Chinese relative the American paddlefish's snout has been thoroughly examined, revealing thousands of ampullae of Lorenzini running in two rows; those same electrical sensory cells that sharks have but in significantly denser concentrations.

American Paddlefish, Mermet Springs, Illinois, USA- ***Daniel Zemans***
American Paddlefish filterfeeding- ***Daniel Zemans***

They inform the fish not just of plankton and other prey but potential obstacles like branches and river banks. Its brain combines the data from each individual sensor to create a virtual map of its surroundings, allowing it to pinpoint a meal while its competitors are still clambering around in the dark. For infants the paddle is essential in helping them track down solitary invertebrates, especially after heavy rain stirs up the water. Any juveniles that lose their paddle, something easily snapped off by an attacking bird, will perish soon after. But in adult paddlefish the loss of a paddle, providing the fish survives the injury, appears to have little impact on the fish. The difference in survival rate may result from the adults targeting plankton in such large shoals that they can find them without relying wholly on their unusual organ.

American paddlefish have a feeding technique all of their own. As they approach a potential meal they start waving their bodies and paddles at it, driving the plankton towards their mouth. Then they do what appears anatomically impossible. Like a shark a significant proportion of their skeleton is comprised of cartilage and is highly flexible. As the paddlefish begins to opens its mouth the lower half of its head slides back to allow the jaws to balloon out in front of it. The bottom jaw drops effortlessly, sucking in the rich glut of plankton while the sudden stream of water out the gills forces the food down the fish's throat. All of this happens in just a matter of seconds. Its Chinese relative had a slight improvement on this technique, not only did its bottom jaw drop but it could also be protruded forward during the process. The paddle in both these species aids in funnelling the plankton into the mouth.

The American species has to travel widely to feed as it soon depletes any plankton it finds. Like its former cousin they still need to return to the same area each year to breed; their most observed spawning grounds being in the Osage River in Missouri. The eggs stick easily to any rocks or logs they wash against, where the young develop safely until the day they hatch and are washed downstream. When the young mature they too will make the return journey to the spawning grounds. But this migration may be different to the ones their parents undertook. People are as prone to affect the rivers in North America as they are near the Yangtze and once again a species of paddlefish is threatened by dams it simply cannot circumvent. In Canada the paddlefish has already been declared extinct in the Great Lakes where it once roamed. At least in the United States populations of American paddlefish are being bred in captivity, as a safe guard against any threats to their wild relatives.

This captive breeding contributed to one of the most remarkable and unexpected evolutionary developments in modern science. For, while the American paddlefish were looking comparatively secure, the same could not be said for their distant sturgeon relatives, including the critically endangered diamond sturgeon (a species with a distribution from Eastern Europe, to the Caspian Sea and north to Russia), and the paddlefish offered a potential solution for that. There was a hope that the paddlefish milt was similar enough in structure to sturgeon sperm to trigger development of the sturgeon caviar in artificial conditions while the vast genetic difference would be enough that any resulting young would only have their mother's genetic material- they would essentially be clones. It was a gamble; pushing science and nature to its very limit. But somehow it seemed to work. Tiny sturgeons later wriggled out of their artificially inseminated eggs, awkward and gawky and very much alive.

The youngsters were born hunters, eye-balling a potential target and speeding straight towards it. But some of the youngsters were doing things a little bit differently, their snouts weren't the pointed weapons of their siblings but were markedly blunter- requiring a slightly different feeding technique. They had paddles. They were the hybrid species that was not believed possible, one deserving of its own name, the Sturddlefish. And there were several hundred of them.

The paddle development, and obvious genetic influence from their father, was variable; some sturddlefish had no paddles at all and appeared to be clones of their mothers- as they were meant to, others had a complete paddle and very few obvious sturgeon features. Some sturgeon characteristics were dominant, such as diet. Regardless of mouth shape they all inherited a carnivorous diet; even those infants with the most paddlefish appearance were not able to filterfeed. Perhaps this was the original diet of the common ancestor of both paddlefish and sturgeonfish. That ancestor lived about 160 million years ago, significantly long ago that any living progeny between its distant surviving descendants shouldn't be able to exist; the sturddlefish is living proof against that. But only for the short term. The scientists involved in their care were determined for no more sturddlefish to be bred; a new hybrid was not the conservation assistance that the sturgeon mothers needed. Nor are the sturddlefish likely to be able to replace themselves naturally, they are most likely sterile. It is not known how long a sturddlefish life expectancy is, perhaps a decade or two, perhaps much longer. About one hundred survived their first year of life, their most vulnerable, so these survivors are continuing despite their unique inheritance. These unexpected offspring are not ready to give up all their secrets just yet.

Saddled Bichir- *Diegograndi*
Poll's Bichir, Mbomo, Republic of the Congo- *John P. Friel*

In West Africa the reedfish could scarcely be more different from the paddlefish, or the sturgeons for that matter. These are small snake shaped fish and just like reptiles they continually have to surface in order to pop their head above water. In the tropical areas where they live annual floods can wash enough mud and silt into the water that it begins to clog and stagnate. There is oxygen here but it is above the surface of the water, mixed in with all the other free floating molecules that make up the air. To survive here reedfish have to find an alternate method of breathing to gills and they have one. Sharks improved their breathing efficiency by reducing the number of gills and modifying one of the redundant gill arches into jaws that could gulp in more water. Reedfish have modified this design again. Only four of their gills are used in the typical way, the fifth gill arch has, over millions of years, been modified to form a purpose built pouch that swells when filled with air and is rich in blood cells to soak up the oxygen. So, if it feels short of breath the reedfish only has to lift its head and fill its primitive lung with air.

The reedfish bears a strong resemblance to juveniles of another species of fish in the same area, the bichir. There are eleven species of bichirs and when young they too have a thin body that stretches out like a skipping rope. But as the juvenile bichir grows it develops other features that make it a more substantial predator. The biggest difference, literally, is the size between the two groups; the very largest reedfish only measured 37cm, some bichir can be over a metre in length. Bichir also have an additional support in movement; pelvic fins, positioned roughly where the lower limbs would be in a land vertebrate, which aid the fish in negotiating the tangled labyrinth that characterises so many swamps.

It is when the wet season arrives in August that the bichir really heats up the action. Bichirs of several different kinds follow the deluge into new territory. Cuvier's bichir, a large grey species, are one of the most conspicuous, not due to their colour but because of their behaviour. A solitary female hovering below the surface is most obvious at first but careful observation will soon reveal a male cautiously trailing her. He is not alone for long. Another soon joins him and the courting male forgets his original target and turns on his rival. The pair sizes each other up and if neither backs down they start to compete against one another in the most dramatic way. The males begin to leap as far out of the water as they can, crashing violently back into the swamp. Even though bichir can breathe air this still requires a tremendous feat of energy. The resulting splashing attracts other admirers. Fish eagles fly in towards the disturbance, talons outstretched to catch the obvious targets. Why do the males do it? There is a risk of course, but it allows them to vividly demonstrate their strength both to each other and the discerning female. All of the fish involved in this epic contest are unaware how much of their lifestyle is due to their primitive lung.

Their rudimentary organ provides another advantage to the bichirs and reedfish. A critical element for a hunter in these swamps is being buoyant enough to float effortlessly below the surface of the water. The cartilage of a shark's body is only moderately heavier than the supporting water causing them little difficulty with this. But the bichir, like all other ray finned fish, has solid calcium bones giving them a specific gravity twice that of water. Sterlets and sturgeons simply resolve to retire on the river bed while paddlefish have to keep continually swimming to avoid a similar fate. By coming to the surface to swallow oxygen the bichir inadvertently fills its body with enough gas that it raises its buoyancy.

Continually travelling to the surface to achieve adequate buoyancy is time consuming particularly for fish that don't live in seasonally tainted waters. So most modern bony fish have modified what is in the bichir a lung into another specialised organ, the swim bladder. Like a lung the swim bladder concentrates oxygen so that a fish can maintain its position in the water. To change depth however requires control over this organ. When the fish wants to ascend its red blood cells release more of their oxygen into the swim bladder. To sink the fish's body simply transfers oxygen back into the blood cells so that their effect is diluted and spread around the rest of body. In this elegant way these fish have achieved complete buoyancy control without being tied to a need continually return to the surface. But a consequence of this specialism is that if one of these advanced fish is trapped in a stagnant lagoon most of them are no longer able to breathe the fresh air above. They have sacrificed this anatomical life-raft for efficiency in the water. Some of the earliest fish to deploy a swim bladder can be still be found today in North America, the gars.

The most primitive of the existing gars is, like the paddlefish, a giant. Not only is the alligator gar nearly three metres long it has a broad solid beak for snatching prey. Like all gars this giant has restrictions in hunting. Its top jaw is attached solidly to its skull and hence can't be moved, restricting its ability both to chew and the extent to which it can open its mouth. Nevertheless gars can still hunt with great efficiency by spearing fish between their jaws. The alligator gar has two rows of teeth in its top jaw so that when it closes its mouth they jut out below the top lip, like the reptile it is named after. It resembles the alligator in more than just appearance. They are skilled predators, unleashing an unexpected violence to make a successful kill. They have even been known to attack and eat alligators a metre and a half in length. The alligator gar is unusual among its family for another reason; its swim bladder is poorly developed making this the only living gar that can breathe surface air, allowing it to thrive in estuaries and muddy swamps

Spotted Gar- *Diegograndi*
Eyetail Bowfin, Huron River, Michigan, USA- *BelowBlueWaterDiver*

All other gars are restricted to cleaner waters but the payoff is the increased buoyancy control their advanced swim bladder allows them. The spotted gar is typical of the group. Because of its reliance purely on its gills for breathing it only reaches a third of its cousin's size. The jaws of this gar are slim and delicate, quite a contrast from the robust design of the alligator gar. Most vertebrates have nostrils near the base of the skull, but not the gar. It has nostrils at the very tip of its snout, a rather vulnerable position but one that allows the fish quicker access to passing scents. Spotted gars are designed for ambush and wait near the surface, ready to spring on passing fish. The gar shoots out after its prey like a spear with such speed that it occasionally catches birds and drags them down to suffocate them before eating. But resting at the top of the water brings its own risks. Here the gars are clearly exposed to their own predators. They have significant protection. Their body is covered with hardened scales, coated with an enamel like substance. Most predators would struggle to break through it, and while they're taking the time to do so would be vulnerable to attack from the gar itself. There was one hunter this body armour did not protect them from however. Humans. Their admiration for these scales led them to use them as arrow heads, as the tips of these scales easily pierced into prey. There is a peculiar irony in the defences of such a highly effective killer being used as an equally effective means of attack- including, at times, against other gars.

In the same area a close relative of the gars is a rather more active hunter. Not only is the bowfin larger than nearly all gars, excepting the alligator gar, it has a more solid build. Instead of a pointed beak it has a broad head, not unlike that of a buffalo, that it uses to crush prey rather than spear it. A stocky streamlined body combines with smooth overlapping scales to reduce the level of drag and make it a swift hunter. Bowfins also have a gentler side.

In April male bowfins begin to congregate in the shallows around the river. The overgrown vegetation here impedes their movement so it is harder to successfully hunt. But the decaying leaves and branches have another purpose. Gathering them in his mobile mouth the bowfin begins to drag the debris into position around a submerged tree trunk. With no fingers or limbs to assist him the bowfin relies on his jaws alone to arrange the material into a sphere. He carries over small stones to anchor the construction into place. He is making a nest. It will have to be of the highest standard as it will house his young until they reach maturity. Males become territorial; driving away any animals, or curious humans, that may threaten his work.

All this careful work eventually attracts the attention of a female bowfin. The male may be more colourful but she is significantly larger. Nonetheless after all the labour he has expended on the nest the male has no intention of letting any intruder near, including potential mates. He begins to nip the female as she approaches, continually harassing her. With the initial assault she may retreat a little way. But she returns continually until the male gets used to her. By this stage both are ready to spawn and the female is finally allowed to approach the nest. The male bowfin is so picky because after mating he will be left alone to guard the nest. This takes a constant toll on him and some bowfins relax their typical aggressiveness to share the nest with another male. Once the eggs have been laid the focus is no longer on competition but on protecting the next generation. There is a host of creatures from alligators to even snails that wouldn't mind supplementing their meal with a few rich fish eggs. The bowfin charges at the smaller threats until they leave his territory but for hunters like alligator that would equally enjoy snacking on him he has few options but to remain motionless and take his chances.

Things hardly improve once the infants hatch. The bowfin father keeps a close eye on the infants milling around him but even so he is likely to lose at least a few to hungry predators. In the open they rapidly vanish but there is one place he knows they are safe; the nest that he has constructed himself. So he steadfastly shepherds them back to it over the coming weeks. Only once the juveniles are old enough to leave home will he resume seeking out enough prey to sustain him till the next breeding season. Their behaviour is not the only advanced feature of the bowfins. Unlike other primitive fish the bowfin can move its top jaw. This may sound like a simple development but it provides critical advantage when it comes to biting and swallowing. It also allows the bowfin to chew its food and digest it more thoroughly. All modern bony fish have taken advantage of this mobile jaw. Of the 35,000 species of ray finned fish more than 99.8% have this quality. Together this group is called the teleosts.

In America are found the most primitive of living teleosts, the mooneyes and goldeyes. These closely related fish reach at most half a metre in length and visually appear relatively unremarkable fish. They look like completely modern fish, until they open their mouth. The position of the upper row of teeth is standard enough but not the lower. Scientists, when they decided upon a scientific name for this fish gave it the rather fitting name of *Hiodon*, tongue teeth. The lower jaw has teeth also, but it is that tongue that provides most of the biting power.

Mooneyes are also unusual for living in the northern hemisphere. But primitive teleosts originally evolved, and for the most part stayed, in the southern hemisphere- spreading from South America to Australia and South Asia. Those most closely related to the mooneye have further developed the skeletal reinforcement around the tongue and consequently are called bony tongues. They have moveable top jaws but about half of their biting power results from the tooth lined tongue. To consume its meal the bony tongue presses its unique organ against the roof of its mouth to slice its food into pieces. The bottom jaw acts as little more than a convenient place to protect and house the tongue.

One group of bony tongues, the featherbacks, typically shelter among weeds in the murky darkness. Several different kinds exist across Asia, some in India, others in Thailand and Indonesia. The clown featherback is a metre long but its dorsal fin is tiny; projecting over its back like a feather. Decorating the fish is a series of large dark spots, each surrounded with a neat white border. It is distinctly hunchbacked and has an undefined tail which simply tapers away. The reason for this peculiar shape can only be found by examining the fish internally. The body is filled by an elaborate swim bladder which stretches nearly to the tip of the tail, and overflows into other cavities throughout the body, distorting its shape in the process. But this swim bladder does not have the internal lining of other species that use the swim bladder as a lung. It is likewise too large to be used purely for buoyancy control. What could it possibly be used for?

The featherback hunts at night like a bat, echoing sound through its vast organ. Its swim bladder has been adapted as the brass instrument of the marine world. But it has a second equally important purpose. Producing sound, while efficient for navigation, can easily be heard and avoided by small fish and tadpoles. So the featherback uses a separate detection technique. The large swim bladder is sensitive not just to changes in pressure but to the small electrical charges emitted from all living creatures. Its convoluted organ has become the featherback's equivalent of those electrically sensitive cells lining the face of a sturgeon. The featherback uses its equipment in the same way, to locate suitable prey before attacking. Having such a reduced dorsal fin reduces the rebound effect on the fish. But the clown featherback's way of finding food is primitive when compared to that of its sole African relative. This is the smallest and one of the oddest of the featherbacks, the false featherfin. This little brown fish has lost its dorsal fin completely resulting in the least interference of any featherback on its sensory system. The strategy has obviously paid off as the false featherfin is one of the most common fish darting among the rotting vegetation. But even it is outclassed by another group of African fish.

The mormyrids are the largest family of bony tongue fish and all species use electricity to hunt. Unlike sharks or sturgeons they don't have either sensory pits or barbells to pick up vibrations. Instead they produce their own electricity, continually emitting little beats, typically about eighty, which resound around the fish. If anything interrupts the electrical pulses then the fish draws closer to investigate, moving back and forward over an intriguing spot. If it is nothing more than a stick then the inanimate object will only impede the pulses and will soon be ignored. But if the disturbance is caused by a live fish or snail then the mormyrid will be able to detect the tiny release of electricity caused by the animal's beating heart. Even the most expertly camouflaged and motionless prey will stand little chance of avoiding detection.

The dorsal band whale is a small mormyrid and engages in searching out food socially. This fish is pretty indistinct from the myriad of other fish hunting in the Congo except for its distinctive movement as it wriggles back and forth seeking prey. The dorsal band whale can only detect surface prey but other mormyrids have advanced the technique. The sneaky bulldog shares a similar body to the dorsal band whale except for its snout. Jutting from the lower jaw like a crude finger is an unusual appendage. It resembles a barbell but this organ detects electricity not taste. As it hunts the sneaky bulldog leaves just this organ dipping into the substrate. It increases its perception of the charged environment around it to include that hiding below the sand.

The presence of other electrical fish jars the signal so the sneaky bulldog drives away any of its companions that draw too near. To have success finding food it must locate it alone. The elephantnose fish is even more sensitive to locating a hidden meal. Its sensor organ projecting from its chin is so long that it resembles an elephant's trunk, hence its common name. Due to its size and the sensitivity of its equipment it too is a solitary hunter but one marked by its unusual intelligence.

Worm-jawed Mormyrid- ***African-Parks, Chinko, Central African Republic & Teleos***
Sneaky Bulldog, Ukuzwana-Witbad nature reserve, Mkondo, South Africa- ***Pieter Kotze***

Elephantnose fish, Aquarium St Petersburg, Russia- *Aleksandr Levon*

The brain that allows the elephant nose to prioritise between the multiple electrical signals is enormous. Not only does it have the largest brain to body weight ration of any fish it is also comparatively larger than all reptiles, birds and most mammals. Even the large brains of monkeys and apes are petite by elephantnose standards. So are our own; on average our brains make up 2% of our body weight, in these tiny fish it can be up to 3%. Their large brain is exceptionally energy hungry- sixty percent of the oxygen used up by their body is consumed in the brain. The development of its brain is most obvious in the cerebellum, where some of the most complex neural connections of any animal are formed, although it lacks a cortex- the outside section of the cerebellum where most of our more complex thinking and behaviour is generated. People use our complex brains to communicate and control our social behaviours allowing us to form intense relationships with hundreds of people. But, to the best of our knowledge, this intelligent fish uses its skills for feeding alone and has no tolerance of companions, or their interfering electrical signals.

Yet even elephantnose fish have to come together to mate. The ideal condition for the young is after the monsoon hits; when the temporary increase of habitat can support their growing numbers. In the darkness and mud the fish cannot hear the falling rain or even feel it falling with their sensitive lateral lines. But they can sense the arrival of the wet season in a way that we can only imagine. The current of electricity is significantly affected by water; it is diluted by it as effectively as a handful of salt into the Amazon. As the water level rises the electrical currents produced by the fish are absorbed quicker. They begin to tolerate the presence of other elephant nose, until they allow another close enough to mate. That is the end of their relationship, neither bothers to guard the eggs and if any of these develop and reach adulthood it will have achieved this entirely without any assistance from its parents.

Other electrical fish are gregarious long term. In Africa nyanda fish hide during the day in submerged caves in places such as Lake Malawi. Many of the local cichlids could outpace the hunters but at night things are different. As the nyanda leave their caves something unusual about them becomes clear. Several venture out together; this is no typical solitary electrical fish. Juvenile nyanda would have little chance competing directing against gangs of adults and they sneak off on their own to feed on shrimp and other such prey. If they are lucky they will overturn a freshwater crab and feed on its unarmoured legs. A couple times a year they may also come across blackfly larvae gathering in their millions before they leave the lake. But for the most part juvenile nyanda remain safe at the edges of the lake.

The adult nyanda fish are rather more adventurous. They hunt across the lake in groups which range from a pair up to ten. In packs any larger than this the fish seem to be too numerous to detect their electrical senses properly. The pack's noisy presence drives their prey together. Cichlids, the most common fish in the lake, can't detect electricity but their lateral lines feel the combined movement of so many hunters and they can certainly hear the boisterous predators. As the group draws near cichlid try to escape, foolishly often heading straight into open water. Lake Malawi is no ordinary lake; it consists of an area of thirty thousand square kilometres. While sheltering near the caves and plants of the lake the cichlid's electrical signals are somewhat muffled by these small obstructions. Once in the open the strong muscular movements of the cichlids quickly expose them to the electrically sensitive hunters. The nyanda pack swarms together but their feeding is far from social. They will snatch at prey as soon as it draws near whether it is rightfully theirs or not. They will also steal prey that has already been caught by one of their companions.

With such fierce competition it is understandable that some adult nyanda try to seek out prey on their own like the juveniles. Without the protection of a group they are soon driven away to the least productive parts of the lake to feed. They may pass a sleeping cichlid during their hunt but without the group disturbance the prey is unlikely to be scared into the open. Despite the occasional loss of a meal hunting in a pack appears to be more effective than feeding solo. Such a form of social hunting appears to be confined to the nyanda fish.

South of the Sahara one bony tongue, the aba, has become a supreme solo hunter like the sturgeons of the north. It could almost be mistaken for a conger eel if it wasn't for its almost complete absence of fins. The aba uses only its tail and a single pair of fins behind its gills to move forward, keeping its spine rigid as it swims in a manner peculiar to itself. It keeps stiff for a reason. Like the rest of its family it hunts sensing electrical stimuli from its prey. But unlike the elephant nose the aba doesn't rely on a trunk like projection on its jaw, but on the dense muscles that run across its main body. Muscles are naturally reactive to electricity; it is via the firing of small zaps of electricity into the nerves that any animal is capable of moving at all. But moving these muscles would automatically overwhelm this hunter's ability to pick up the small electrical traces of other animals. For the aba to detect anything it has to keep the muscles of its main body completely rigid, requiring a method of swimming that doesn't rely on either its tail or its chest fins. Consequently the aba's dorsal fin, unlike any of its relatives, stretches across most of its back, raised up like a vast sail. Through wriggling its dorsal fin continually the aba is able to swim, freeing up its abdomen to find the next meal uninterrupted by the rest of its body.

Although the adult aba is adapted to living in the muddy river bottom the foul waters will not sustain its eggs or developing young. To breed aba must head to the surface. They rip away overhanging leaves to construct a floating nest about a metre across. This nest will hold the eggs, preventing them from sinking to their destruction. Such a nest draws the attention of numerous other fish looking for an easy meal and the parents have to guard it aggressively to ensure least some of their young survive. Once the eggs begin hatching the aba determine the period of protection over and plunge back into the darkness.

In West Africa another bony tongue also comes to the surface to breed. Unlike the Aba, the transparent eggs of the butterfly fish are naturally buoyant enough not to require a nest and they are abandoned. The parents also hang around the surface; this is a bony tongue that never goes grubbing around in the soil. Its vertical pointing mouth shows that it plucks up prey from above, in the process taking a few gulps of air.

Butterfly fish go to the surface primarily not to breathe but to hunt, mainly insects. Most insects can fly though. This is not a problem for the butterfly fish. Its compact body is cloaked by enormous fins which protrude from its chest. As it moves towards its prey it flicks the fins down with such force that its body leaves the water. Once airborne it holds the pectoral fins motionless, like a hang-glider, as it soars towards its meal. Unlike an insect the butterfly fish is incapable of true flight; its fins are designed for gliding not flapping. Still so perfectly does it achieve this glide that it could be mistaken on first sight for a bird. They have been doing it for a long time too; even longer than scientists were expecting. A DNA analysis in 2011 comparing populations between the Congo and Niger River revealed that while physically they are nearly identical genetically they have been separated for some fifty million years.

Sharing the waters is another distant relative, but one considerably larger- stretching to about eighty centimetres or so. But it is no threat to the butterfly fish. Its name, the African bony tongue, doesn't shed much light on this fish as the elephant nose, nyanda and false featherfin are also bony tongue fish and all live in Africa. There is another feature about it that is unique to this group; its diet. Just as in the Mississippi and Yangtze rivers at times there are vast plankton blooms here. This plankton attracts predatory water bugs and small fish, wriggling through the green bloom. Any other bonytongue would be only interested in the other animals but this unusual species is interested in the lot; snatching up great mouthfuls of the plankton in the process. Indeed, it's the only species in the group that feeds extensively on plankton. It will also slide between the rotting debris below the reeds targeting individual minute prey. Sometimes this docile fish will plunge its head into the mud itself, filtering out any edible morsels in the process, even swallowing small seeds.

Butterflyfish, Aquarium Poland- *Łukasz Olszewski*

The African bony tongue also consumes less edible fare, sand, using it in its gut to grind down its meal the same way a caged budgie does a piece of cuttlefish. Due to its flexible diet it can afford to grow large, easily dwarfing its African cousins. It can also occur in large numbers and sustain local fisheries. In Benin alone, a West African country neighbouring Nigeria, the value of their harvest was estimated to be worth 1.5 million in a single year. Unlike other ancient fish, like the sturgeon, the African bony tongue is, so far, keeping a stable population or even increasing due to its use in aquaculture. It has even been introduced into other regions, like Madagascar, where it has started to threaten local species. The ancient African bony tongue is proving surprisingly successful in an ever changing world.

Despite its common name the African bony tongue's closest relatives don't actually live in Africa. Instead they live in South America. It used to be thought that, just as the ancestors of sturgeons spread out across the northern hemisphere, that the ancient bony tongued fish conquered the southern hemisphere in the same way. Recent discoveries have radically disrupted this theory. The restriction of bony tongue fish to the southern continents is a comparatively recent development; fossils have been found in the Northern hemisphere, as far away as Greenland.

Right across Europe multiple different types have been found in the eras following the extinction of the dinosaurs. Some of these unexpected rebels to one of science's pet theories were even identified as marine species. These marine fossils are particularly important. It is now considered, using the current range of fossils and DNA analysis available to us, that the surviving bony tongue fish must have descended from marine ancestors at least four times for the current variety of known species (including fossils) to have occurred. For some unknown reason all the marine species died out. So too did all those in Europe, Northern Asia and North America, perhaps outcompeted by the sturgeons or other fish species, or perhaps pushed out by the increasingly cooler temperatures of the higher latitudes. Perhaps it was a combination. But those elsewhere, particularly in the tropics, continued to thrive.

In South America the bony tongue fish are not as varied; there are no elephantnose or butterfly fish to be found here. But there are considerable giants. The pirarucu or arapaima, the African bony tongue's closest relative, competes with the giant sturgeons and Chinese paddlefish for the much disputed title of largest freshwater fish. A fully grown pirarucu may reach two hundred kilograms and a length of two and a half metres. Although similar in shape to its African cousin it is distinguished by its oversized scales and its bright red stripes. With the exception of its pectoral fins all of its fins are placed far back on their body. Slyly it can slip through flooded forest, occasionally popping its head above the water to take a breath. While the swim bladder aids in maintaining buoyancy the pirarucu still uses it primarily as a lung and the heavy concentration of blood vessels lining it improve its efficiency. Coming to the surface also provides opportunity for snatching up small prey like lizards on overhanging branches for this is no vegetarian. Nor does plankton have any interest to this giant, only any small fish that may be caught feeding on it. Even a drinking bird may be swallowed in an easy mouthful.

The pirarucu is as mysterious as it is enormous. At one time there was only considered to be a single species spread out throughout South American but studies are now indicating that there are likely at least five species, each previously restricted to their own portion of the continent- although the casual introduction of pirarucu around the continent prior to this discovery means that many now live in locations previously occupied by other species.

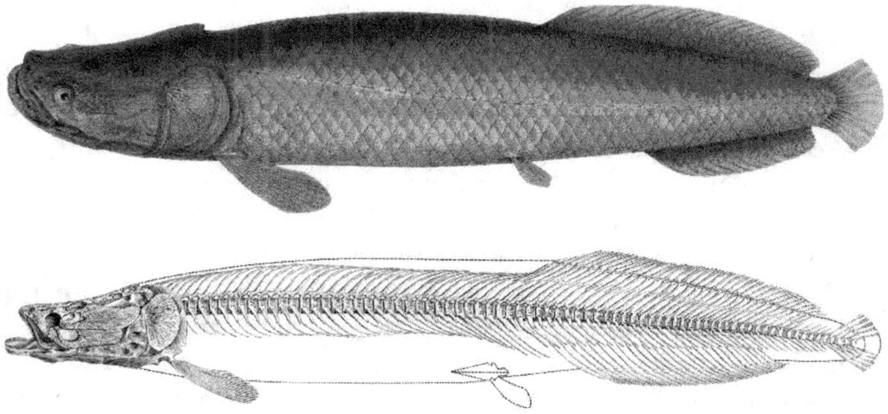

Agassiz's Arapaima -*Spix, J.B.V. and L. Agassiz*
Pirarucu/ Giant Arapaima, Amazon River, Iquitos, Peru - *Manuel V. Miranda del Solar. @manimiranda (instagram).*

One of these newly discovered pirarucu, Agassiz' Arapaima, was considered likely extinct by the time it was recognised.

Agassiz' Arapaima was only known from a single Brazilian specimen collected in the early 1800s and even that specimen had been lost during the bombing of Munich during World War II, although not before records and at least one sketch was made of it. This rarity did have some unusual feature, but not ones that could be casually noticed; it had, for instance, differences in the numbers of teeth than its relatives. Experts worldwide started closely examining whatever arapaima they could find, not just in its original habitat but in aquariums, fisheries, tourist attractions across the world; indeed pirarucu have been so successful in India that they are considered an invasive species. Examinations of those multiple individuals helped to establish the five different species, and suggests that there is a likelihood that Agassiz' Arapaima still exists today, potentially already captured in someone's holiday snaps, or swimming misidentified as one of its cousins in an aquarium somewhere.

The very habitat the pirarucu lives in is part of the reason for the initial divergence of the different species. As the Amazon and other major river systems enter the dry season some rivers may be cut off from each other for years- long enough for the restricted populations to gradually change. When drought is severe enough fish might be caught within rapidly diminishing pools. Pirarucu are able to continue taking in oxygen through both mouth and gills as the water stagnates. But this isn't enough. The mud itself hardens around the large fish. The pirarucu begins to dig deeper into the desiccating swamp seeking any remaining moisture. As its gills clog up it begins to rely increasingly on its primitive lung for all its oxygen needs. The pirarucu has transformed the swim bladder from a buoyancy accessory back into an essential component of its survival, a survival technique that it also shares with its smaller African cousin.

The pirarucu can outlast the dry season but it is only when the Amazon overfills once more that it can bring forth new life. Unlike the bowfin both parents share the task of protecting the next generation. A pair creates a nest among the submerged plants, plucking away bothersome branches and leaves with their formidable mouths. Once they are content with their progress the female lays 47,000 eggs inside the shelter. The parents divide the labour of defending the nest as effectively as they did constructing it. By taking turns guarding both parents are able to leave the home territory to feed and avoid the starvation that a single parent would otherwise have to endure.

Watching sedate eggs may be arduous but it is a lot more energy-consuming once the precious parcels finally hatch. Young fish are prone to stray but the pirarucu has means to keep them near, chemical means. Glands in the head produce pheromone laden mucus that wafts around. The substance attracts its offspring but its effectiveness depends upon the size of the young. While newly hatched they are intoxicated by the pheromones, even nibbling the mucus on their parent's forehead to obtain more of it. As they grow their size and tolerance increases in turn allowing them to swim further away from their guardians until they take on the solitary lifestyle of the non-breeding adults.

The pirarucu is not the only species of bony tongue in these waters. There are also arowana; moderately sized, stockier, slightly distant relatives. There are two fairly similar species in the Amazon, the silver and the black arowana, the colour difference being about the only thing that splits the two apart. Being shorter than the pirarucu makes it easier for the arowana to chase down prey. Two barbells project from the lower jaw like a fork, picking up any scents floating in the water. A shrimp waving its legs around a scrap of algae catches its attention. Slowly the fish draws close enough that the invertebrate comes in contact with one of the barbells. Its lower jaw suddenly drops open, like a hidden trap door, sucking the shrimp into the throat and snapping shut with an equal speed. Like its giant relative the wet season is boon time for the arowana, particularly due to the countless small animals stranded among isolated trees. Some like butterflies or grasshoppers can fly or at least leap part of the distance from one plant to another. But most, whether ants, scorpions or amphibians have to enter the water. Once immersed they all become easy pickings for the arowana. But the arowana's diet is not restricted to animals in the water. It also investigates leaves hanging just out of reach for suitable prey.

Katydids, a sort of rainforest cricket, tend to frequent these highly risky areas, relying primarily on their camouflage for protection. If all else fails they can hop out of reach of any approaching frogs or birds. But they are not as alert of threats from below them. With eyes set high on the forehead the arowana can see the katydid well before the insect has any indication that it is there. In an instant the fish has surged forward, dragging the leaf and its dislodged occupant below. Sometimes, though, the prey is too high to be caught by just popping your head above water.

Silver Arowana- *Statuska*

Several feet above the rising water level are intricate spider webs but the arachnids within them are not an easy catch. Changes in pressure on the strands of its web quickly alert its inhabitant if the supporting branches are becoming submerged. When this happens the spider simply consumes its construction and rebuilds it higher in the canopy. To reach its relentlessly rising prey the arowana has no option, it has to jump. It takes advantage of its considerable body size as it lunges through the water to catch the miniature hunter. If it needs to it can lift almost its entire body out into the open air, but this visual spectacle last just a few seconds.

Adult arowana are mighty predators but their eggs and young are just as vulnerable as any other small fish in the Amazon. They could easily be swallowed by a passing caiman, pirarucu or even a turtle. So the silver arowana takes his offspring into his own mouth. Only the male protects the young in this way, as the female is already developing a new brood of eggs inside her. A pair of jaws, even one as large as an arowana's will not hold as many eggs as a nest so this fish does not invest in thousands of eggs but less than two hundred. Small parents may only have about seventy. As each egg is about a centimetre in diameter they soon cram the mouth to capacity.

While still in the egg the fish are sustained by yolk but once they hatch they need to feed and there is little opportunity for this inside their father's mouth. What was once a safe haven has become a prison, with little chance of escape. But the arowana parents have prepared for this also. Each baby hatches with an orange piece of yolk attached, dangling from its abdomen like some kind of parasitic jelly bean. This will sustain the juvenile until it is large enough to control its buoyancy and swim properly. Only now will their dad's enormous lower jaw drop open and allow the infants to hunt for their own food. But even then the father stays near and the shoal of burgeoning young can dart back to the safety of an open mouth if any potential threat approaches. The infants can continue using the shelter of their father's mouth for sixty days after hatching until they can no longer easily squeeze inside their protector's jaws, or indeed any other fish in the Amazon.

Almost identical relatives of the arowanas are found in the East, so similar that many are called Asian arowanas. They are so successful that they are the only bony tongue fish to be found naturally in the northern hemisphere today, living as far north as Vietnam. Each region has their own distinct patterns and there is some debate as to how many species there actually are, although most scientists consider there to be only one.

Borneo alone has three different types of Asian arowana; one with a grey body and a pink tail, another with a silver body and slightly redder fins, and a third species which is a rather distinct chilli red colour- all three living in their own separate part of the island. What nearly all Asian arowanas share are their solid body, large reptilian eyes, large round scales that look like they could be substituted for body armour if they had to, powerful pectoral fins, and a pair of barbells that flicker in front of its lower lip reminiscent of a medieval monster. Across much of South East Asia they are more frequently known, not as the Asian arowana, but as the dragonfish.

The fresh water dragonfish has become associated with beliefs as mysterious as its namesake and was collected in vast numbers as a good luck charm. Consequently few of them are left in the wild and due to their significant size many captive individuals patrol tanks on their own with little chance of ever using their advanced parental techniques.

An Australian relative, the saratoga, is still common, and its comparatively dull appearance means that those few which have been smuggled overseas can be recognised comparatively easily.

Dragonfish- ***Sommai***
Northern Saratoga, Crystal Creek, Mapoon, Queensland-
Alastair Freeman

Across North eastern Australia the two species of saratoga inhabit some of the country's great river systems. There are predators here that could make short work of a baby bony tongue and the parents are equally mindful of their young. In saratogas and the dragonfish the female takes on complete care of their young. After mating she gathers the eggs in her mouth and begins the solitary venture of parenthood.

In the billabongs a pair of saratoga can easily be separated when the water level drops so by holding onto her young the female can ensure their survival by guarding them constantly from spawning through to adolescence. Indeed, she will not tolerate having any other saratoga around, including the father. Once the young are old enough, and the rains return, they are unlikely to have any more to do with their mother for long; those that are unable to find their own patch of billabong will soon find themselves targeted by their previously much devoted protector. The famished mother saratoga, is finally free to return to her normal level of hunting activity, unburdened by a mouthful of young, until she is recovered enough to bring another generation into the world.

The bony tongues may be careful parents but they were only the start of the vast teleost group that would soon take over not just the fresh water of the world but the oceans. These teleosts could rely on their heritage of a swim bladder to maintain buoyancy as effectively as their competitors the sharks. They had one significant change to their bony tongue relatives; no longer would they need to use a tooth studded tongue as a substitute jaw; their lower jaw would do the job for them. Not only did this allow them to be free of the expense of growing surplus teeth, most importantly, their jaws were now able to able increasingly flexible shapes to specialise further, allowing them, in time, to be able to consume an increasing variety of diets, and live ever more varied lifestyles. This seemingly insignificant change provided the tools to allow them to become the most numerous of all the vertebrates.

Chapter 4. Swarming Shoals

Tarpon- ***TravelSync27***

As the tide enters the mangrove swamps of a Caribbean island something prehistoric glides in with it. Juvenile fish take shelter from it amongst the roots of the trees but this will be of no avail. Tracking them is a hunter that has survived the extinction event that killed the dinosaurs, ammonites and every type of aquatic reptile except for turtles and crocodiles. Ten percent of the ocean's bony fish were also destroyed; the survivors were basically the most advanced of the teleosts; the vast group of fish defined by their mobile mouths and powerful symmetrical tails. Yet somehow one ancient teleost species has survived virtually unchanged until the present day.

The tarpon is one of the most enduring of all marine teleosts. Fossils show that the tarpon, or very similar looking ancestors, have been terrorising the ocean for over 100 million years. The tarpon still has traits that link it with the earliest of all teleost groups, the primitive bony tongues, which becomes obvious when it surfaces to breath. Like the bichir it continues using its swim bladder as a sort of primitive lung. This ancient feature still has its advantages.

As schools of juvenile fish become trapped in the mangroves by the falling tide the stagnant water begins to take effect. The trees holding this turbid swamp together also trap layers of mud sealed over rotting vegetation. As it dries the increase in temperature causes gas to seep out of the debris. Even above water the smell is overwhelming, resembling that of an overflowing sewerage drain. Underwater the foul bubbling is literally suffocating. Small fish may be able to survive momentarily in these conditions but only by reducing their energy levels to next to nothing. The tarpon does not have any such limitation. Its mouth opens like bellows, easily sucking in the airborne oxygen above it. Such a mouth can be used equally efficiently against its prey. The hunter dives below its stranded victims then begins to rise, panicking the juveniles to shoal closer together. As it makes its ascent it keeps its mouth sealed to the last moment, reducing the amount of drag. When it reaches them the tarpon suddenly stretches its jaws wide open, taking advantage of the suction generated by the movement to inhale several small fry in this one swift movement.

Tarpon do not begin life as such efficient hunters. Infant tarpon look different enough from their parents to warrant their own name, leptocephali. On first hatching they resemble a crumbled leaf, while by early adolescence their body stretches so that they resemble worms, not particularly different in fact to the larvae of a lamprey. Leptocephali are not unique to tarpon. A number of its close relatives, like the ten pounders and ladyfish also start life as insignificant wriggling larvae. Another relative, the bonefish, can be found like tarpon close to the coast. They tend to congregate around sandy beaches, sucking crustaceans and snails into their mouths. There is another group of relatives that are equally as fierce hunters as the tarpon. But it would be hard to guess it from their origins.

Floating, somewhere in the Sargasso Sea, are several million eggs. Their parents died shortly after spawning and their decaying bodies have helped to fuel the growth of plankton and invertebrates that await the hatching young. The leptocephali here, almost from birth, begin to split into two distinct groups. One will begin to swim relentlessly to the coast of North America. The other will travel on to Europe but its journey is a slovenly one at that. Most of its migration is spent chasing food, snatching up passing plankton. In this way a trip that would have taken only ten months if it had simply drifted with the current will take eighteen months if not more. As it reaches land the leptocephali begins a dramatic transformation as its body stretches and it takes on a light pink colour. Shoals of the larvae begin to wriggle up sluice pipes and small waterfalls in their way. If an obstacle is too large to be passed they will wait until night and leave the water altogether; sliding through wet grass and other vegetation, and relying on the surrounding moisture to keep their gills damp enough for them to still breathe, until they reach another stream. They continue relentlessly upriver. By this stage they are quite clearly elvers, juvenile eels. Their cousins in North America have also developed enough to be recognised as freshwater eels, although there is still debate as to whether they are the same species as the one in Europe.

The male European eels, after swimming hundreds of kilometres inland, stop temporarily and begin to harvest whatever sustenance they can find. The females continue on until they may even reach the Alps. The females grow much larger than the males so it is likely that if they remained at lower altitudes with lower levels of oxygen in the water they would suffocate. Once again, like the basking sharks, it seems to be the gills guiding at least part of this migration. All this time the females continue feeding and growing until eventually they reach an altitude high enough to temporarily pause their journey and begin to gorge themselves.

After eight or ten years the females begin to make their way back down to the sea. The males, lower down, may only wait three years to resume their journey. As they enter increasingly saline water their skin becomes darker until they enter water so deep that you cannot distinguish them anyway. Eels' sensitive lateral lines distinguish disturbances in the water, they may also detect mild changes in gravitational pull and magnetic waves allowing them to navigate across vast stretches of ocean, although there is still debate as to how much each of these different cues actually guide the eels. Eels may also potentially be able to see at least part of their journey due to one final change in their body. In land vertebrates birth marks the end of development of the eye and its maximum growth. But the development of an eel's eyes at this later stage of its life is closer to that of a baby still in a womb. Consequently their eyes are able to swell out in front of their heads, leading them straight on their course. They are making their way back to the Sargasso Sea to spawn for the first and only time of their life. That done they will all die. That the eels can survive at all is due only to the versatility of body shapes in the developing animal. An elver simply could not survive in the open ocean nor could the leaf like leptocephali last for long if it was transplanted directly into a fresh water stream. A fish like the hagfish, which hatches as an identical miniature of its parent, would not be able to occupy such different habitats in one lifetime. But because of the versatility of forms the eel goes through in its life they are able to not only survive but thrive. By the time the miniature larva develops into an adult it could reach a metre long in some of the larger European females.

There are freshwater eels in the southern hemisphere that grow even larger. In the island of New Guinea and some of the nearby Pacific Islands a spectacularly large eel can be found swimming through many of the rivers. Due to its immense size the long fin eel has obtained much spiritual respect and associated customs from the local people. A tribe may gather to feed the eels at the water's edge which eagerly awaits the free meal. It easily dwarfs any child brave enough to swim with it. Some monster eels are even larger than the human parents that cautiously watch their children play with the giant fish. Most of the fish are about two metres long or so but some individuals that may get stranded on their way back to the sea, by a dried up stream for instance, may reach three metres. Such size can only be appreciated out of water and the local people lure such fish closer and closer to the edge.

Eventually someone dangles the food partially onto dry land and the eel follows it up, lunging its body half out of the river to reach it. The emerging eel easily snatch the food out of the person's hand, although there is always the chance with such a large animal that it may bite one of the fingers offering its meal in the process. It is essential that the eel is quick sliding back into the water or it may strand itself and not be able to return. With the day's entertainment over the people return to their forest and the eel continues searching for other suitable prey. Cannibalising on younger eels provides one way to maintain their considerably bulk. They will even try the occasional duck.

Growing to such a size does not happen in a few months. It has been estimated that some of the larger eels may be thirty five years or so. Being so large is important for, like its European relative, the long fin gets just one chance to breed. If anything their journey is even more perilous than their northern cousins. They travel south to the Coral Sea, aiming for a location near New Caledonia. The tropical water here is savaged by constant storms and in summer hurricanes may form, dragging currents and struggling fish with them. Any weak or injured fish are quickly lost early on their migration.

Marbled Eel, New Ireland, Papua New Guinea- ***Tom Vierus***
Migrating American Eel elvers, Halifax, Nova Scotia, Canada- ***Hunter Stevens***

Upon entering the ocean the long fin, for the first time in many decades, finds itself dwarfed by the surrounding water and other fish. It has no option but continue on. Eels from right across the Eastern Pacific are heading to the same stretch of sea.

Just as in the northern hemisphere several species of eel seem to share the same spawning area. Some have come from the opposite direction, all the way from the southern and eastern coasts of Australia. In their millions they squeeze round each other, in a marine ballet that is yet to be observed by any human. But each player in this aquatic drama meets a fateful end. Eels have no mechanism to control the development of their sex cells, so all of them, regardless if they're eggs or sperm, mature at the same time. In mammals or sharks, by comparison, once an individual reaches reproductive age a small proportion of its sex cells mature in cycles throughout the animal's life. But in the eels, of both hemispheres, after they have bred once, that is it. Their bodies have put everything into the production of their successors and are exhausted. With that done, the eels perish, sinking to the depths, creating an annual feast for the denizens of the deep, some of which may in turn be eaten by the developing larvae.

Other eels spend their entire life in salt water. There are a number of different families; one of the largest is the snake eels. Unlike the long fins few snake eels live long enough to reach a considerable size so they have to rely on other defence mechanisms. Their only fin is the tail making it easier for them to dive under the sand for cover, without chest or pelvic or even back fins getting in the road. While several species spend much of their life sheltering in this way most of the family swim conspicuously at midlevel and need other survival techniques. They are swift when compared to river eels but they are still burdened with a body little different to a skipping rope and easily outpaced by the efficient body of a tarpon or jack. So snake eels have found another way to reduce the risk of attack; imitation. Sea snakes patrol the coasts, easily defended with their venom filled fangs. They are the most poisonous of all snakes, rivalling even cobra and taipan; in barely a minute one bite can stop an attacker's heart. Bright stripes of colour circling their body warn of its armoury, as conspicuously as the markings on a skunk. The harlequin snake eel isn't venomous but it closely resembles a local sea snake; with its long black body marked by vivid white stripes. Other snake eels imitate even more colourful species of marine snake and again the impersonation is a close one. The eel is masquerading in fact for its life; its survival rests on a hunter being fooled enough by its appearance not to test it further.

There is a group of eels that can rival even the snake eels for colour, the morays. For the most part morays stay hidden safely in holes in the coral with just their head popping out. And what a head it is. In the Caribbean a vivid pink species may be found, in Africa the leopard moray is soft blue with black spots and in Hawaii the dragon moray is not only an elegant visual mismatch of creams and yellows but has nose tubes so elongated they stand up like horns. Perhaps the most extraordinarily coloured is the Australian ribbon eel. The male is a spectacular blue with a little fan shaped flap of skin hanging above his nose. As he ages though he changes gender and colour; eventually morphing into a fluoro yellow female. Juvenile ribbon eels stand out even further with their ebony bodies and yellow dorsal fins. Most other reef fish are brightly coloured, ironically, so that they can blend in with the surrounding reef but the morays flaunt their colours to such a degree that this is clearly not their purpose. Instead the eels want to be seen by potential predators. Many morays have high concentrations of poison in their skin and their gaudy colour acts as a warning sign as potent as that of a sea snake; 'don't eat me, I'm dangerous'. If they feel threatened they will stretch their entire colourful body out of their shelter and puff up aggressively. If this is not enough the moray can defend itself equally well with a mouth full of teeth.

A moray often lies with just its head out of its shelter gaping wildly. This isn't just about aggression; in order to keep its large body supplied with oxygen the moray has to rely on its open mouth as well as its gills, constantly sucking water in with its formidable jaws. There is more to their bite than can be seen with the unguided eye. After snatching prey into their mouth most fish rely on suction to take this food into their gut. To do this requires using their mouth like a vacuum cleaner, gaping to draw water deep into their throat to push the food the last part of the way. But morays have a peculiarly narrow mouth, so narrow that the only animals they could catch by this method would be minute copepods and other small shrimp, hardly enough to support such a large predator. But the moray's main diet is medium fish, which they swallow with surprising ease. Their secret is hidden deep inside their throat.

Many rayfinned fish have a peculiar extra feature, a second pair of jaws Many rayfinned fish have a peculiar extra feature, a second pair of jaws, but the moray eels use their additional pair in an unparalleled way. This second pair typically lie flat and out of the way. But as the moray grabs a damselfish in its mouth this hidden pair of jaws, springs into action. The extra jaws pin into the struggling damselfish and slowly begin to drag it down the throat. As the meal is swallowed the eel slowly loosens its grip on the damselfish with its primary jaws and the only evidence of it (and that extra pair of jaws) is a brief expansion of the throat. All of this happens in just a matter of seconds. The moray can pop its head safely back inside its hole while it digests its meal.

There are some meals that will lure even a moray eel out of its hole. A triggerfish is a large and relatively tasty meal. It isn't poisonous but it does have one significant defence; halfway down its back is a muscular rod which it can raise at will. Most of the time it lies flat but when a predator draws near the triggerfish uses the rod to wedge itself securely within the coral. Any shark that tries to pull the fish out will only succeed in lodging it tighter against the surrounding rocks. But the moray does not act like a shark. Gently the smooth hunter grips the triggerfish in its mouth and slowly begins to rotate. The triggerfish begins to wriggle closer to the rocks using its full strength to stay rooted to the spot. But it is to no avail. The second set of jaws locks into gear. They secure the prey, leaving the main jaws free to drag the fish deeper into its mouth. The moray begins to twist the trigger around, despite the fish's vigilant shaking as it tries to stay in place. Eventually the moray dislodges the trigger enough that the entire fish slides out of its shelter and into the eel's waiting mouth.

Morays may be seen as fearsome, lonesome hunters but there are some things that even an independent eel can't solve by itself. One need in particular causes the morays trouble, oral care. In an attack pieces of flesh often rip off prey and snag around the moray's teeth. Over a couple of days the scraps begins to rot, drawing bacteria into the eel's open jaws. But rotting food doesn't last stuck to the moray's teeth for long. The fish has its own specialised dental team. Several different species of shrimp attend to it, indicating their nature to the eel with their stylised movement as they approach, allowing them to be distinguished as not just an unusually obliging meal. The shrimps use their delicate bodies to work their way between the narrow grooves of the moray's teeth and over its sensitive tongue. In these moments the eels show an unusually calm nature as their attendees work on them.

Mediterranean Moray Eel, Cabo Cope Puntas del Calnegre Regional Park, Spain- ***Alberto Carrera***
Co-operative hunting between Chestnut Morays and Leopard Groupers, Cabo National Park, Mexico- ***Pat Webster @underwaterpat***

Recently another, rather more unexpected companion, has been observed. Although fast enough to dart after an unwelcome snorkeler and give them a decent bite a moray's speed, for the most part is easily outpaced by most other fish- at least in the day time. Consequently most species, such as the chestnut moray, spend most of the day time half-buried within its shelter. There is one creature that can draw them out however, a grouper. Groupers are completely unrelated to eels and are more than large enough to pose a reasonable threat to morays, indeed smaller size morays are at risk of being preyed on by these substantially sized predators. But in this instance the grouper is not interested in the moray for food but for something else entirely. It positions itself vertically above the burrow and purposefully waves its head to side from side until it has enough of the moray's attention for the eel to follow it to better hunting grounds. Some groupers are more direct and harass the reluctant moray until it rushes out of its burrow after them.

Despite its speed the grouper has a serious limitation when it comes to hunting; most of its prey is slim and agile enough to hide within coral or other forms of shelter where the hunter is incapable of reaching them. The morays however can take full opportunity of their ability to squeeze within narrow spaces- and any fish that try to escape their jaws by fleeing into the open risk disappearing into the grouper's open maw. The fish that both predators target are for the most part small enough to be swallowed in one go- reducing the direct level of competition between the two species and allowing it to be so successful that sometimes large shoals of the groupers and multiple morays can form; hunting with efficiency that few of their prey are capable of escaping.

Of all the eels none has inspired more folk legends and superstitions than the green moray. Worldwide the green moray is well known for its aggression and its tendency of chasing overfriendly swimmers out of its territory, frequently leaving them with a parting bite on their behind in the process. There are actually nearly a hundred different species of green moray, most of them sharing the same territorial traits, barely leaving its sheltering hole during its life. But some unusual individuals decide to venture beyond, not just its shelter, but the reef itself. These enterprising adventurers travel to the depths of the ocean, not to breed but to feed. Their usual live prey is rare here so they have to join sharks and other scavengers fighting over dead fish on the sea bed.

There are other eels that venture into such depths regularly, conger eels. Several individuals live permanently without any available shelter and have to rely on their size and aggression to survive. Such fish are rarely seen unless a substantial meal, like the rotting carcass of a tuna that has sunken to the deep, attracts them. Not all conger eels are such solitary hunters. One small family of specialised congers, garden eels, feed not on other fish but on plankton filtered from their surroundings. Compared to other eels they have two main problems. Firstly, they have an exceptionally slim build; even the largest ones, which may reach a metre, have a body-plan which is essentially that of an oversized noodle. Secondly, they have very small teeth so they cannot even bite predators in defence. On the reef itself there are many competitors for plankton and eels simply don't have the speed to compete so they are forced to eat on the sandy unprotected perimeter of the reef; where their slender shape and defenceless mouths make them an obvious target for predators. If they can catch them.

Garden eels have found a way to create their own shelter, they dig. Wriggling their tail vigorously back and forward they manage to create tunnels with ease. As they ease into their hole they produce a slime which prevents their tunnel from being washed in. To feed they simply pop their head out of their burrow and pluck off individual plankton as it passes by. Their main concern is the location of their burrow; it needs to be in a relatively bare area where constant currents bring a regular flow of food right past them. Such areas, like a sand bar, are often in short supply and where they occur the garden eels often give up the solitary habits more typical of eels and tunnel close to each other. They do maintain a territory, of sorts; each vigorously protecting the space immediately surrounding their burrow. Living at such quarters also simplifies breeding significantly. To breed a garden eel simply has to itself half out of its tunnel until it comes into contact with an eel of the opposite sex. Once a pair forms they wrap their upper trunks around each other, keeping their tails safely in their tunnels. In this rather awkward position they spawn, darting safely back below the sand the moment they've concluded, while their fertilized eggs drift perilously with the current until they, if they are highly fortunate, happen to come across a sand bank similar to their parents.

There is one other major advantage to living in groups. The garden eels may have shelter to hide in but this is only effective if they have enough warning to retreat to it. Living on its own an eel can rely only on its own senses to avoid danger. A simple slip of concentration while feeding could leave enough time for it to be attacked and consumed. But communal living means that as soon as one garden eel darts for cover its companions soon notice and copy the movement. So highly co-ordinated is their overall movement that the image of the garden eels swaying smoothly with the current was enough to remind explorers of terrestrial plants blowing in the breeze and resulted in their name. At times their choreographed behaviour looks almost comical but if a threat draws near it could also save their lives.

Spotted Garden Eel- *Dmitry Rukhlenko*

Sandlances also shelter under the sand when predators draw near. These fish are relatives of a completely unrelated group from eels, the perch or spiny ray finned fish family, but their behaviour sheds light on how a comparatively solitary burrowing fish might have been transformed into a more social shoaling species. In low tide they stay safe in the sand but as the tide returns they begin to venture further out and start gathering their fill of plankton and other small invertebrates.

These fish are slim, small and defenceless. They have no poison, no deceptive colouration or additional set of jaws to defend themselves with. But they do have numbers in their favour. A few dozen set out together, continually changing position as each tries to stay at the secure centre of the group. They may stick tightly together but each harvests food separately; it is simply a shoal of convenience. The fish watch each other keenly, reacting swiftly when they see any change in direction that might imply an incoming threat.

The sandlances' communication is aided by something shared by nearly all fish, a lateral line. These are specialised groups of cells that run across either side of the animal, detecting any movement in the surrounding water. These lateral lines can quickly pick up the indications of predators and also allows it to sense and coordinate its movement with the rest of the school. It is critical for the survival of most shoaling fish. So it might be surprising that the most common and successful of the shoaling fish, the clupeoids, namely the sardines, anchovies and herring, have lost their lateral line entirely.

There is one primitive member of the clupeoids order, the denticle herring, which we have already encountered. Unusually among the group it lives in fresh water. It is also the only species that still has a lateral line. When rain washes mud into the rivers it relies on this inbuilt sensor to avoid running into the surrounding banks or rotting vegetation. Assumedly it must also assist in avoiding predators, allowing the denticle herring to seek out some form of protective cover in the darkness.

In the open ocean conditions are better but even in the clearest water you're lucky if you can see further than two hundred metres. A great white shark at top speed can cross that distance in less than thirteen seconds. And it's not the only predator here; sailfish, marlin, tuna are all looking for an easy meal but unlike fresh water there are no tree roots or large rocks for potential prey to hide behind. For a timid plankton feeder like the Atlantic menhaden the only available shelter is each other. The fish draw close to each other as they play a risky game. Their safest option is to stay at the centre of the shoal, where they will only be eaten if the entire school is lost. But while the menhaden stay here they slowly starve, as the hordes of mouths protecting them inadvertently also get the best picking of the surrounding plankton. To feed properly can only be done on the high risk fringes. It is a gamble that menhaden and all shoaling fish must play each day. Staying together in groups filled with fish of similar age and size makes competition fairer between members but it also reduces the ability of predators to distinguish one individual to another. A larger, or younger, menhaden would be easier to focus on and would soon be snatched up.

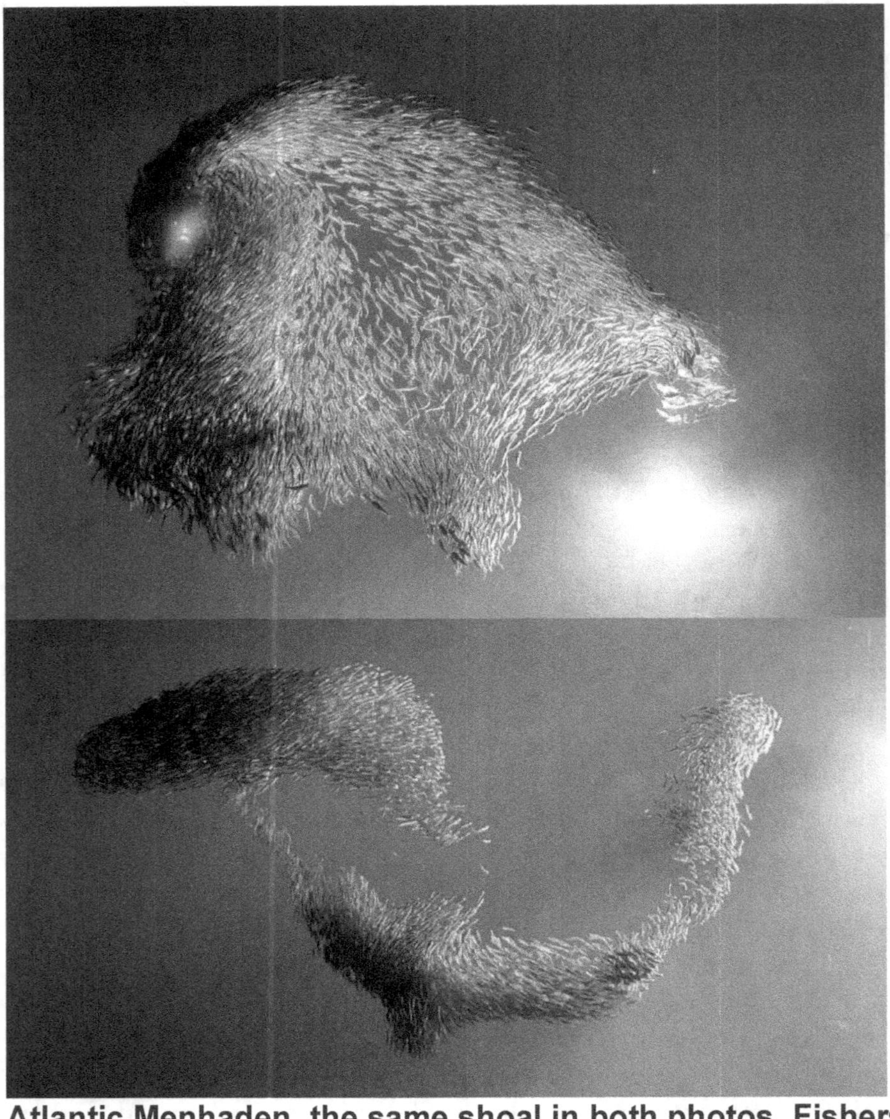

Atlantic Menhaden, the same shoal in both photos, Fishers Island Sound, New York- *Murray Fisher*

Another shoaling species, the anchoveta have another secret to their survival hidden in plain sight upon their skin. Like most shoaling fish, its body is silver and like a well-polished car it reflects any incoming light. As a fish moves, curling its body through the ocean, this reflection continually gleams off its scales. In a single anchoveta this effect is mesmerising, in a shoal it can be breathtaking. To the anchoveta this visual feast is also instructive.

As each fish moves its companions can easily see what direction it is heading and follow it. Anchoveta are perfect mimics, following one another's movements fin stroke for fin stroke. A few individuals shimmering to one side can set the entire group moving in a new direction. The anchoveta use this technique not only to move the shoal into new feeding grounds but also to avoid potential danger. They have a great need for this protection.

When a lone seal approaches the shoal it only takes the movements of a single fish to trigger the entire group to take evasive action. Uniquely anchoveta have a highly developed rostrum organ that projects out with the upper jaw. This aids the anchoveta in smelling out a threat well before it sees it. It is now that the anchoveta have to rely on their copy-cat skills to save their life. A seal foolish enough to dive through the middle of a shoal will find it parting around its hungry mouth. Several seals may have a little more success. By breaking up the anchoveta into smaller groups they can finally manage to catch a few. Survival for these fish comes down to numbers, the larger the swirling mass the more intimidating the shoal become. In their most dramatic performances the shoal forms an impenetrable circle. In such an immense group a predator is almost thwarted in picking out prey. Just as soon as it has visually singled out one fish its target has disappeared back into the endless shoal. But, impressive as they are, the anchoveta do not form the largest shoals in the ocean.

Pacific pilchards, or sardines, also gather in their millions. This species is so widespread that it can be found in Chile, Australia, Africa, California and Japan. So spread out is the Pacific pilchard that it was for many years considered to be five different species. They spend most of their time out at sea feeding on plankton. But every year or so cool currents are drawn up against the coast of South Africa. Plankton blooms in these conditions and the pilchards surge in after them. Juvenile pilchards are usually sold as sardines and it is the sheer abundance of these juveniles that often lead to the entire group being called a sardine shoal- regardless of whether it primarily consists of adults or juveniles. Sardines exhibit the gregarious nature of clupeoids more dramatically than any other fish. A single shoal can contain more than a million pilchards. Several billion pilchards may travel north with the cool waters of the Benguela Current. When conditions are right these sardine shoals form the biggest gathering of marine life on the planet.

But the sardines are also trapped by the cold water. They cannot leave it to feed or even escape hungry predators. Sometimes the current will turn and pin the shoal against a projecting coastline, helpless to flee from the army of dolphins, sharks and seals assailing them until the current moves on. Only when the Benguela Current begins to withdraw back south are the sardines drawn back with it, where they can shelter in the deep water and begin to rebuild their numbers. In the near freezing waters of the north another relative also faces the challenges of being penned in by the temperature.

Across the North Sea and North Atlantic as winter draws near immense storms rip up the ocean, mixing up plankton and nutrients from deep below but destroying any exposed animals in the process. As temperatures begin to drop ice forms temporarily on the surface of the sea only to later melt and sink; adding to the mayhem of the winter drama but also displacing more nutrient rich deep water in the process. Cousins of the sardines, the herring, are natural inhabitants of the icy water but even they have to seek shelter. Off Norway herring begin to travel into deeper waters, taking advantage of the natural fjords that penetrate so deep that they are well protected from the storm raging above. Eight billion herring converge together, waiting for the sun to return.

Even in the gloom herring still have to keep watch for predators, killer whales and cod track them and come to harvest what they can. Like the rest of the clupeoids herrings have no lateral line and vision is little assistance with no light to guide it. But herrings do have an additional and peculiar form of communication, flatulence. Herrings use this excess gas to keep in contact with each other, but its exact purpose is unknown. It may be used as a warning signal of predators or perhaps as part of dominance displays between individuals. It could possibly be even used to avoid crashing into each other; what little that is known about their communication in captivity is that the more herring there are, and the darker it is, the more gas each individual produces.

When spring arrives and the storms finally settle the herring are free to leave the fjords and start to feast. For while the storms raged the nutrients they whipped up from the deep began to fertilize the microscopic algae in the plankton. The seas turn green as result and the herring emerge to gulp up and harvest the easy meal. Herring do this not just in the Atlantic but in the Pacific and their main priority soon turns from feeding to breeding.

Pacific Herring eggs covering beach & close-up, Seattle, Washington, USA- ***Sara Roolf***

Off the coast of Alaska herrings travel in their millions towards the shore. Herring are picky about where they lay their eggs ensuring they will not be washed up or destroyed during fiercer weather. The fish dart around plants where their rows of eggs entangle like loose strings of cotton. So many eggs are deposited in a few hours that massive patches of water are transformed into a milky white colour, large enough to be spotted from a passing plane. The water froths with so much eggs and milt floating around that any fish swimming in it soon disappear from view. The sheer amount of eggs may seem excessive but it is very unlikely that many will last long enough to hatch, let alone for any of the fish within to reach adulthood. Nonetheless enough herring survive each year for this extravagant strategy to be successful.

Further north other fish are also emerging in their millions from their winter shelters. Arctic cod might vaguely resemble herring but they are not related to them or the sardines or indeed any other clupeoids. Cod belong to a completely different group, a super-order containing 1,400 different species. They have broader, more solid bodies than herring but their most obvious difference is the single barbell that dangles like a hook beneath the chin. This is a sensory organ that it can wriggle through the substrate if it has to to detect hiding prey. Cod have other sensory adaptations to help them hunt; unlike the clupeoids, but like most other fish, they still have a lateral line to pick up movements in the water around them. While other fish are sheltering from the worst of the winter weather the cod venture out as supreme hunters.

The surrounding waters reach freezing point yet somehow the Arctic cod can not only stay alive but continue actively swimming. Inside the fish's body an array of modified proteins act as an anti-freeze, keeping the blood inside the cod liquid enough that it can continue to flow around the animal. With this adaptation cod numbers are not limited by how many can squeeze into fjords for shelter and populations are able to survive winter with minimal losses, despite the frigid temperatures. Great shoals move under the ice, some containing 500 million thrashing fish. Its Atlantic cousin can also survive low temperatures, allowing the population to keep feeding throughout the winter and thrive as a result. A century or so ago this cod could have challenged even the sardine for the title of the world's most numerous shoal but overfishing has changed that. Even so under the right conditions their numbers can once again become vast.

Like the herring the Atlantic cod take advantage of the right weather conditions for breeding. Due to their large size they reproduce at their youngest at four years old but they are not fully mature until they are six. Being so well adapted to the cold they breed in winter, the large females soon attracting the closest males to attempt to breed with them. The males begin to jostle fiercely with each other, fighting desperately to get as close to the female as possible. Eventually the smaller, immature or sickly males are driven away so only the largest, and healthiest, males are close to the female when she releases eggs. The largest females, about one and a quarter metres, can produce more than nine million eggs. But intensive fishing pressure removed most of these larger animals. The smaller cod survived by slipping through the holes in the net. Fish stocks began to crash extensively beginning in Iceland and Norway then spreading to Canada. In 1992 Canada imposed a ban on harvesting cod in the Grand Banks. With such fertile animals the cod were thought capable of soon recovering their numbers. But decades later although fishing has been able to somewhat resume the cod are yet to completely recover. Instead with the high population of undersized or immature males few of a female's eggs were now being fertilized. In turn small females of just a metre in length, can only produce about 3 million eggs. There is also a likelihood that many of these may prove infertile. The age segregated shoals that their own offspring rely on for protection are subsequently significantly reduced in size and more vulnerable to predators. Mortality is already naturally high amongst newborn cod but all of these additional impacts mean that the cod were barely maintaining their population let alone increasing it. It may take many more years until vast shoals of countless cod once more migrate through the frozen ocean.

A month after the Atlantic cod spawns a smaller relative begins its courtship. Haddock are also highly competitive when it comes to choosing a mate and they have their own special way of deciding which is most dominant. Two males may surround a female, circling her vigorously so that she will not stray too far. Like many other fish they use their swim bladder to control their position in the water, but they can also use it for more delicate purposes. They contract the muscles around the swim bladder to produce a sound that echoes around them. The males continue beating their organ like a drum, clearly beating out their dominance, singing with the only voices available to them. The pair increases their volume and speed until one finally retreats exhausted.

The remaining male then, rather more gently, begins to court the female with the same instrument. In the process he wryly flashes his fins, moving gracefully closer to his potential mate. Only in the moment of spawning will the male finally stop his singing, completing his work in silence. So noisy are the haddock that they are occasionally heard above water. But, loud as this is, some of their relatives are even more vocal when it comes to creating new life.

Toadfish are ground dwelling squatters that could scarcely look more distant from the cod and haddock but they are related. The splendid toadfish of Central America are unusually gaudy with their striped pink bodies and lemon fins. But this tropical species is the exception among its family and even its bright colours are only so it can blend in with the coral. The male plainfin midshipman from Alaska to Mexico is much more inconspicuous during the day with his pale brown body. He hides for most of the day in a burrow he may have constructed using a boulder, a clam shell or even a soda can as one of the sides. But when the mating season draws near this toadfish finally begins to announce his presence. He waits until nightfall then begins to call, endlessly, utilizing his swim bladder to send his message resounding for kilometres through the water. He does so at such a volume that the males can be heard above water, loudly enough to wake nearby human inhabitants.

In the Californian town of Sausalito, it did not take long after the local waterways were cleaned up for this mysterious noise to start blasting each night. Scientists were sent to investigate, as it was so loud that Russian submarines were even considered as a potential cause. Upon discovering the noise pollution was caused by a singing fish the town decided to hold a festival, which kept going annually for the next couple of years, clearly deciding that if you can't sleep you may as well party. The male midshipman certainly has other things than sleep on his mind and the continual racket soon draws the attention of a female. To help him stand out even more he has rows of luminescent organs running across his flanks. As these organs glow they provide a lantern which the female can use to guide herself with. Once the female reaches the booming male he briefly stops his call long enough to grasp her between his fins and mate. The female deposits less than a hundred large eggs which stick to a suitable surface beside the burrow by means of a sticky disk, then vanishes. Other the next couple of weeks other females may arrive, so that eggs of all ages end up pressed tightly together.

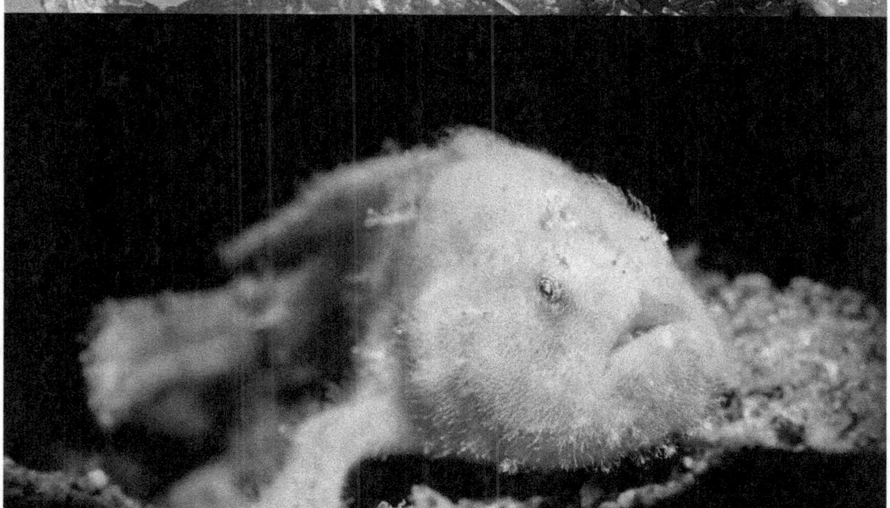

Plainfin midshipman father egg guarding, Shilshole Bay, Seattle, Washington, USA- ***Virgil Daggert***
Striate Anglerfish, Bali- ***Joe Belanger***

Such a collection would provide a welcome banquet to any passing starfish or octopus so the nesting males don't leave their vigil during the breeding season.

The protective fathers will only hunt what they can snatch up directly in front of their nest. Consequently they quickly lose weight compared to their non breeding bachelor companions. Even if the eggs become marooned from him when the tide drops he will not leave them. When the young hatch the father continues guarding behaviour and may even shepherd shrimps and other small prey towards the infants for them to practice hunting on. After a few weeks the young will need to fend for themselves as the male midshipman resumes hunting for his primary diet of small fish. After his fast any infants that overstay their welcome may be too tempting to be ignored. It will be another year before the Western coasts of North America begin to hum once more to the sound of toadfish.

Monkfish lead more inconspicuous lives and lack any vivid lighting or other colouration that might set them apart. For the most part they are drab fish that huddle on the seafloor waiting for prey. Their body is covered with furry filaments that allow them to blend effortlessly into their surroundings. Their most distinctive feature is an isolated portion of their dorsal fin that flickers above the fish's mouth. This fin moves in a mesmerising pattern that is attractive not only to people but to any small fish wandering these waters. Should they move closer to investigate the monkfish will show an impressive feat of energy, leaping forward to gobble up the fish before returning to its sedate existence. Both the head and jaws are unusually wide allowing the monkfish to swallow just about anything within reach. And it's not just fish that are fooled by the fish's lure. Diving birds, including famously a goose, have all found their way into this predator's stomach. So effective and widespread is this lure waving technique that the family to which the monkfish belongs is fittingly called the anglerfish.

One of these anglerfish is often found hiding in the floating algae of the Sargasso Sea, where it feeds on the larvae of passing eels as well as other prey. The sargassum frogfish lives just as easily in other floating material, whether that be drifting palm branches or even discarded fishing rope or nets. Its mottled colouration allows it to blend in smoothly with most backgrounds but it also has some limited control of the pigments of its skin, allowing it to choose which to suppress and which to display in order to further improve camouflage. The pelvic fins of this fish have been modified so that they can grasp, like miniature hands, allowing it to cling onto kelp and swing gracefully with the same rhythm. So seamlessly does it blend in that unsuspecting ocean dwellers like flying fish may attempt to graze on it.

It is the moment the frogfish has been waiting for and it quickly takes advantage of it. In a movement quicker than your eyes can process the flying fish is down its throat. To discover what has happened you have to slow down time, to a fraction of a second. With the aid of high resolution cameras we can do just that and can finally appreciate the feat achieved by the frogfish. As the meal draws within reach the frogfish flicks its lure behind its head and expands its mouth like a balloon, providing the suction to draw the flying fish towards its jaws. It pushes back onto its pectoral fins, lunging forward in the process. The frogfish completes a rapid transformation, from one of the ocean's laziest hunters into one of the fastest animals on the planet. Indeed it has some of the quickest reflexes in the world, only the snapping shrimps which clicks its claws at prey, matches it for speed. In one six thousandth of a second the frogfish has sprung and secured its catch. Once the meal is safely inside its mouth the sargassum frogfish freezes once again, leaving only its lure swaying to draw in further prey.

Sargassum frogfish have to be vigilant predators to survive among the drifting weed and even cannibalise their own species if they get the chance. Consequently mating between these fish is nearly as perilous as that of black widow spiders. A female will often make the first move towards a male clinging tightly to the foliage. As she draws near he snaps at her, chasing viciously. He may even draw blood. She responds just as brutally, the pair continuing to bicker until they have established equal strength and finally settle down to mate. The eggs are enveloped in a thick layer of jelly and quickly float out of reach of the ravenous parents. Copulation over the frogfish return to their usual stations, before the temptation to attack their temporary partner overwhelms them.

Pearlfish do not have the camouflage skills or the speed of sargassum frogfish. They are slim, transparent and easy food for any that could find them. But being so slim has its advantages; the pearlfish can squeeze into any cracks and crannies it can find around the reef or rocky temperate shores. And the hiding places it can find are surprising.

Wriggling slowly through the sand are sea cucumbers, worm like relatives of the starfish that few creatures are likely to attack. This simple

Fowler's Pearlfish, Bunaken, North Sulawesi, Indonesia- *Andrey Ryanskiy*
Pinhead Pearlfish sheltering in sea cucumber, French Polynesia- *Vetea Liao*

invertebrate has the ability to extrude its internal body contents or even expel poison to ward off threats.

The pearlfish begins to trail the sea cucumber, placing its head close to the creature's gut, listening alertly. If it is satisfied with what it hears the pearlfish moves to the rear of the animal and enters the sea cucumber. Somehow it manages to do this without alarming the invertebrate into releasing its poison.

There are numerous animals that pearlfish will shelter inside including oysters, sea squirts and even starfish; each species having its own favourites. These other lodgings aren't completely adequate as the pearlfish still have to leave to feed and breath. But sea cucumbers have specialised organs called respiratory trees that allow them to breathe through their rectum as well as their mouth so the fish is at no risk of suffocation no matter how long it stays inside the creature. This unusual invertebrate also has the peculiar ability to regrow its internal organs. So any pearlfish that sneaks in here is provided with a free meal as well as housing and adequate air filtering.

Such a prize is rare and explains why the pearlfish checks before entering the sea cucumber. If it makes a mistake and the invertebrate is already occupied the two fish will begin to fight for possession of their host. Such fights will often lead to the death of one if not both of the competitors. Occasionally such aggression may be avoided and several pearlfish may share the same sea cucumber but such gregariousness is rare. For the most part these fish live isolated lives protected from the dangers of the ocean by their unwilling accommodation.

There is at least one occasion when pearlfish must venture into the open however, to breed. Together a group will emerge over several nights and spawn. During the dark several will be lost to predators but the great majority survive. This is not the case for the young, which will spend the first part of their lives exposed in the plankton until they are large enough to enter a sea cucumber. Juvenile pearlfish have their own unusual form of defence. Protruding from their head is a thin spear tipped rod. Depending on the species some of these protrusions are spotted like the fish, while in others they are an opaque colour. These little spikes might appear sharp but those curious enough to reach out and touch it will feel the rod detach and slide softly towards their hand. As it catches the light it can be fully appreciated, a thin filament that floats lifelessly in the current. It is lacking in bright colour, poison, movement or any other obvious way in which it could be even mildly effective in defence. Yet for some reason its unusual shape catches the eye. And that is the basic purpose of the rod. Any predator that swallows it will find little nourishment. All of the worthwhile nutrition remains inside the body of the young pearlfish and it made its exit as soon as it released its rod.

Pearlfish lose the defensive rod as an adult. Having a miniature spear in front of your head would likely prove an encumbrance not only in blocking off part of your vision but when entering a host. Once they are adults they are basically invulnerable inside their shelter, although living inside another animal limits their size. If any were to reach much more than half a metre they would be too large to squeeze inside their host. Exposed outside of their shelter any of these large adults would quickly be picked off by predators. This would reduce the chance of any of these oversized fish producing young and as a result the general size of the whole population of pearlfish is kept fairly stable.

Fish which are more active hunters have different restrictions on their size. Being larger makes it easier to bring down smaller prey but it also requires a more solid build to maintain speed. Becoming bigger also increases the amount of food required in order to fuel their body. For a predator to be large without facing starvation it has to be supremely honed for survival.

The pike family of the northern hemisphere contains one of these supreme hunters. The common species can reach a meter and is a dedicated ambush hunter. When pushed it can put on quite a burst of speed but it is a pace it cannot maintain for long. To save its energy it may hover virtually motionless among a tangle of weeds for so long that you may forget it's there. So will small carp and in their search for aquatic insect larvae they may pass the pike's lair. It is a deadly mistake. In a moment the pike bursts out grasping the carp between its jaws. It shakes the fish desperately, ripping through skin and bones with its sharp teeth as it does so. With its catch confused, severely injured, and likely paralysed the perch proceeds to swallow it.

The muskellunge, another species of pike, is even bigger. An adult almost reaches the length of a man, occasionally getting to 1.6 metres. It is also an ambush hunter and snatches up not just fish but muskrats and voles entering its grounds or even the occasional duck. When attacking such creatures the muskellunge draws up between them and dry land, preventing them escaping out of reach. It then grips the target by the chest and proceeds to rotate it around with its mouth until it can swallow its meal head first. A fish ravenous enough that it will take down a small mammal can barely afford to share food let alone territory.

Muskellunge, James River, Lynchburg, Vancouver- *Chris Crippen*

Consequently, the muskellunge and pike are both fairly solitary, only coming together briefly to breed. But they have smaller relatives that are forced together, like sardines, in vast groups for protection.

In Japan the ayu, a type of smelt, would be an easy mouthful for any pike sized fish. Unlike its larger relative the ayu is not a dominating predator but sustains itself on a rather different diet. Smelts generally gather small aquatic insects but the ayu is unusual in that it primarily grazes on algae, nipping away pieces with its oversized teeth. The ayu has only turned to a vegetarian way of life comparatively recently but it nonetheless demonstrates some of the adaptations that the first browsing vertebrates developed to process their meal. Algae, like most primitive plants, maintains its structure within a dense layer of cellulose. Most fish have a relatively small gut which allows them to digest the remains of living animals rapidly before the meat begins to rot. But plants take longer to break down and consequently the ayu needs an unusually long intestine in order to digest them. Like most fishes the ayu has no stomach and the consequence of its oversized intestine is a slightly bloated appearance. The ayu can find all of its food within the confines of fresh water but some of its more active relatives are forced to spend at least part of their life at sea.

The galaxiids are a family of several dozen or so small fish closely related to the pike and ayu. Most of these Southern Hemisphere fish are so small that they barely catch attention. These are not fierce predators, nor grazers that can take shelter within the algae, so are under constant threat of being swallowed up by any passing hunter as they gather their own meals in the open water.

Although galaxiids come together in shoals generally these are nowhere near as spectacular as that of the sardines or herrings. Except in a few special places like New Zealand where they come together in such numbers that they thrash the surface of the water in their frenzy. A fully grown jollytail galaxias is only twenty centimetres long and is distinctive mainly for its soft blue to olive complexion and the rows of spots on its back. The galaxias start their life developing in the Pacific Ocean. Here there is plenty of plankton and other small prey for the fish to gorge on. While they are young they are relatively safe as sheer force of numbers will ensure that some survive. But as they mature they begin to school off the coast of islands and southern lands like Tasmania, Australia, the Falklands, Argentina and, most dramatically, New Zealand.

The arrival of the galaxias does not go unheeded. Like the fish other animals have over time travelled across the sea to settle in New Zealand. Some like the sea turtles swim in annual migrations to the island. Others, like seagulls flew across the sea thousands of years ago if not earlier and stayed tied to the island. Here they developed into a new species, the kelp gull, larger and more aggressive than most of its relatives. About a thousand years ago they were joined by Polynesian people travelling in their hand carved craft across the sea. Here the Maori people began to enjoy the full wealth that could be harvested from the island, giant moa birds in the mountains, nutritious flightless parrots scrambling through the undergrowth. But the inundation of the galaxias into their rivers must have brought their attention back to the sea. They began the tradition of entering the shallow streams, dragging struggling shoals of fish out with their nets. The population size of the local people was small enough that enough galaxias survived the onslaught to breed for these large shoals to continue well into the late

Jollytail Galaxias, Gisborne, North Island, New Zealand-Peter Hancock

Koaro/ Climbing Galaxias, north of Dunedin, South Island, New Zealand- *Taylor Davies-Colley*

1800s until the impacts of pollution, and introduced species such as trout was enough to severely endanger many populations. So. the galaxias shoals that occur today no longer reach the thrashing schools of yesterday. Groups originally composing a few dozen individuals begin to cluster closer together. Hundreds of jollytails begin to swim together, then thousands as autumn approaches. As the group swells in numbers they begin the most dangerous journey of their lives. Fresh water running from streams into the ocean attracts the galaxias and begins to draw them ever further inland.

Galaxias surge up the trickling streams in small shoals, slipping with ease around any obstacles. In years gone by they would squeeze between the legs of waiting people with ease. Sometimes even when gripped the mucus on their scales lubricated them enough that they could slip out of the grappling hands. Often they'd drop back not, in the water but on the land. Here'd they begin to wriggle back towards the stream before they suffocate. People weren't and aren't the only fierce hunters here. The kelp gulls converge on the galaxias. They quickly clean up any fish stranded among the pebbles as well as any stragglers. The kelp gulls will even fight with the fishermen, dragging away any catch they can steal. But most of the jollytails will have survived the onslaught. They will cease migrating temporarily, continuing feeding until they reach maturity, when they will resume their migration further inland to their breeding sites. Once they have bred, nearly all of them, like the freshwater eels, will die from the exhaustion of their efforts.

The Pacific is the world's largest ocean and in the northern hemisphere relatives of the galaxiids, the salmon, also leave it to breed. After hatching in fresh water the young head into the sea, where they will spend up to eight years developing. Spring for the adults also triggers them to start preparing for the greatest, and last, battle of their lives. They begin to travel into freshwater in their hundreds, back to the waters of their birth. One of the larger species, the chum salmon has an equally large distribution, returning to places as far away as Korea, Alaska, California and Russia. It is in British Columbia where the true homeland of the salmon is and here all five species converge together as they head upstream. Entering fresh water has a number of effects on the salmon. In the oceans the fish has to continually urinate to remove minerals such as salt from its body which would otherwise build up and compromise it. But fresh water has the opposite effect; it drains away many of the key nutrients the salmon needs to survive. So the fish has to change its rate of respiration and waste management in order to keep swimming in these waters. The salmon is one of the few fish that can change almost effortlessly from marine to fresh water but the transition is not complete. The Pacific salmon, although adapt in so many other ways to freshwater, are unable to feed properly here; once they leave the ocean they begin to starve.

The move into fresh water has other effects on the migrants. As the salmon reach sexual maturity they begin to develop features specific to their species, most obvious in the males. Chinooks begin to grow even larger than chum salmon and develop distinctive black spots along their tail. The sockeye, although smaller, is even more recognisable; its upper jaw begins to curl over like a hook, although only subtly in the female. Indeed the whole male's back begins to hunch over and stands out with a pigment as vivid as fresh blood. To contrast that his hooked face turns a steely green, excepting the lower jaw which remains white. The male pink salmon also develops a hunchback but more obvious is the flush of a light pink pigment that surges over his body.

The pink salmon starts tussling with each other soon after entering the rivers and streams. Their migration finishes almost as soon as they enter fresh water for they do not have the strength to overcome waterfalls and other obstacles that lay ahead of it. Because of the shortness of the distance the pink salmon are the youngest of the Pacific fish undertaking their migration, most are not even two years old. The sockeyes' journey is far longer.

The sockeyes have spent four years feeding in the ocean in preparation of the migration. Most Pacific salmon spawn in rivers which wash the newly hatched young straight out to sea. But the sockeye infant remains for a year in the shallow fresh water, safe from sharks until it is large enough to venture into the mighty Pacific. For the sockeye the ideal nursery is a protected lake and the adults will have to swim relentlessly to reach it. Nor will any lake do or even any part of it. Heavy rain overflows the rivers and increases their flow cascading down the mountains to the sea. The fresh water, though tasteless to us, contains microscopic molecules originating from the rivers and lakes deep in the continent.

The salmon smelling the water recognise the distinctive odour of the lake in which they hatched and will guide themselves on with this scent alone. For there are many other lakes the salmon will pass on its

Spawning Pink Salmon, Alaska- *Galyna Andrushko* Chum Salmon migrating, Chilliwack, East of Vancouver, British Columbia, Canada- *Edb3_16*

way, not all of them suitable for developing young, so with only a single chance to breed they will not risk breeding in a new environment.

There is a serious risk in following your way home using the smell from drops of water. After it originally overflowed from the lake the drop rolled effortlessly over sharp boulders, passed hungry grizzly bears with ease and plunged several metres over steep ravines. All that time it was aided by gravity; continually guiding it lower until it reached the rivers that pour into the sea. Water molecules are undamaged by such hazards but sockeyes being creatures of flesh and blood are more vulnerable. To reach the lakes they have to work against gravity, at times crowding into streams so narrow that their backs break the water and they have to continue in single file. Otters and wolves can easily catch them in this situation and even gulls will try to drag them out of the streams.

The most severe hazard is yet to come, great waterfalls separating the salmon dramatically from their breeding grounds. Eels when faced with such hazards can wriggle through the damp undergrowth at night, bypassing the falls completely. But salmon cannot move effectively out of water, they have no option but to attempt to jump the obstacle. At this stage many are exhausted and crash lifeless against the harsh rocks of the waterfall. There are other animals also attracted to the pounding sound of the falls, grizzly bears with their cubs that have been waiting all year for this feast. They position themselves so that the fish have no option but to leap past their open mouths. Sheer volume of numbers means that hundreds of salmon get through, although thousands more have been lost on the way. They have finally reached the breeding lakes.

They have reached their destination but one final challenge remains for the salmon. The females desire to spawn not just in the same lake but the same section of substrate they were born in. Occasionally several siblings return to the lake at the same time both wishing to lay in the exact same area. In a flurry the females' tails send eggs and dirt flying everywhere. The males are also engaged in high octane battling, they are not so concerned about the specific spawning ground just the spawning. Their hooked beaks prove crudely efficient at grappling each other as they fight over the nearest female. So intensely do these sockeyes battle that occasionally a smaller male will sneak around the fighters and mate with the female. These sneaker males may even be so bold as to barge their way between a spawning pair and release their own milt over the eggs, darting away before they are attacked.

But all Pacific salmon, whether male or female, will die after their first spawning. The males will continue fighting amongst each other until no more new females enter the lake and the months of fasting finally catch up with them. The females in the meantime remain watching over their eggs even as they lose the energy to swim and droop helplessly over their nests. They have reached such a level of exhaustion that they begin to succumb to any bacteria or viruses inhabiting the waters. Their eyes glaze over with mucus and the scales begin to crack and flake away. They won't last long. Each year the entire adult population of Pacific salmon perish. The survival of the species now rests on a few scattered shoals of juveniles in the ocean and the billions of eggs lying buried beneath the decaying bodies of their parents.

In the Atlantic other salmon are also heading on an epic migration. Despite their shared name the two different salmon are not as closely related as they sound. Pacific salmon are actually more closed related to rainbow trout than they are to Atlantic salmon. Nonetheless both share the same shape and desire to return inland to breed. Unlike Pacific salmon or the eels the Atlantic female's eggs do not all ripen at once, allowing them to live and breed for longer. Most of the female Atlantic salmon survive their first season and return to the ocean. The following year they will return to the coast. By this stage most of the females will die of exhaustion but some will make the journey two or three times. Not so the males. Nearly all of them will die after their first migration so some have devised another way of increasing their chance of offspring.

At the same time as the adult salmon are swimming up river they pass juveniles making the opposite journey heading out to sea. Most of the juveniles steer clear of the adults to avoid being cannibalised but some of the males are different. They follow the females and sneak into the spawning session. The mature males tear into each other with the same determination of their Pacific cousins, focusing intently on rivals of a similar size. Now the juvenile males' miniature size begins to work to their advantage. Just like the sneaker sockeyes they are small enough that most of the territorial adults ignore them. While the adults continue fighting the juveniles slip in and spread their sperm over the eggs. Being still immature they will be unlikely to fertilize the entire brood but several of the young developing in the eggs will be sired by them.

Atlantic Salmon jumping waterfall, Kuldiga, Latvia-*Akophotography*

With their detour complete they resume their journey to the Atlantic. Through risking cannibalism they have given their genes the best chance of survival.

Salmon are the ocean's most dramatic example of the extremes ray finned fish will go to ensure the existence and survival of the next generation. They along with herring, cod and freshwater eels spend their life as great migrants, constantly travelling in order to feed and breed. But other ray finned fish have specialised so extensively that they remain primarily in one habitat; whether that be as fresh water specialists, blending in on the reef or venturing into the depths of the open ocean. Each of these distinct habitats have their own challenges; a reef while resembling a colourful tranquil rainforest is in reality dominated by predators, changing currents and most of the surrounding 'foliage', the coral, is inedible. The deep sea is as barren as the Sahara desert and much of it is so far away from the warmth of sunlight that the very water itself nearly freezes over. Even freshwater has its challenges and the inhabitants here have had to adapt all sort of techniques from electricity to covering their bodies with taste buds in order to negotiate the often muddy and impenetrable swamps. These entrepreneurs will be the focus of the following chapters.

Chapter 5. Sensory Hunters

Wels catfish, Westerhoven, Netherlands- *Ewout Knoester*

Spring in a Russian lake creates an influx of new life. Halfway across the lake a family of beavers is carrying out repairs on their dam. As the old wood decays it becomes infested with insect larvae. Mayflies drag themselves out of the water and fan out through the woods. Frogs begin courting in their hundreds. Ducks start leading their new families into the water. A few centuries ago these animals would have been at risk of bears or even Amur tigers stalking by the water's edge. Both these predators are long gone. But there is still a danger lurking in the depths. A duckling pauses to peck at some daphnia and the water swirls below its feet. In seconds the duckling is launched out of the air, caught in the mouth of a behemoth catfish. As quick as the fish appeared it is gone. A few feathers are the only indication of what has preceded. The hunter that caught the duck is massive, about two metres long. One part of the fish truly stands out, its massive head. The eyes are so small that the animal is almost blind. Trailing over its head are what guides the predator in the darkness, barbells; six dangling whiskers around the mouth. The stealthy creature is one of the largest catfish, the Wels.

Most of the time the Wels feeds on small carp and especially eels living around the calm waters of the lodge. But mammals can also approach close enough to become potential prey. Water voles snatch away fresh shoots at the side of the lake. Beavers disturb the landscape more dramatically. They cut away trees that go crashing straight into the water, sending vibrations rippling around them in the process. This is what the Wels has been waiting for. Its immense size provides the memento to leap out of the water to pull the rodent towards it. Even so a beaver is not an easy meal. The catfish is forced to rip its prey apart, shaking blood and fur around it as it does so.

The largest Wels catfish recorded was caught in Russia. It was over four metres, although- as is usual with any exceptionally large record holders- it is debatable exactly how accurate those original measurements were. Today overfishing means that monsters of this size are unlikely to be found again; although significantly sized individuals can still be found in Sweden, Kazakhstan, Hungary and several countries in Eastern Europe. Those in Sweden have been isolated so long from other populations that they have unique genetic features, in addition to its notable longevity- up to seventy years. The Wels popularity has also lead to it being introduced into Western Europe. Here, particularly in urban areas, there is little chance in encountering voles and certainly not beavers. There are still water birds and in one reserve the larger catfish were estimated of eating about 200kg worth of birds while their prey were bringing up their chicks, even picking off cygnets- although full grown swans are largely immune. Once the chicks have grown the adaptable catfish turns back to its standard diet of any other fish in the local area.

In more industrialised areas the habitat might be too bare to support waterbirds and the surviving Wels has began to target other potential options. Like pigeons. As the heat of the day builds the birds attempt to take a quick sip from any water source they can before they resume looking for whatever shelter they can find. On the ground they know they are vulnerable, both from cats and stray dogs and any passing birds of prey, so they land in vast flocks seeking comfort in strength in numbers. But the greatest danger for these birds now is not from above but below. The ripples from the mass of sipping birds radiate out into the water cueing the catfish to their presence. With one great big lunge it surges straight out of the water, regularly returning with a pigeon held tightly within its jaws. Curiously this is not a behaviour known in its natural environment. And it seems to be working for them. Growth in their new territories is significantly faster than in their homelands and in parts of Italy more than seventy years have passed since first introduction- more than long enough for some lucky survivors to grow into giants. One recently caught individual is 2.85m long, the longest verified record. It may not be long until we again see Wels the size of which has not been seen for centuries.

Across the world there are other oversized catfish still in existence in their native habitat. In the mighty Amazon goliath catfish patrol the waters of the Amazon and are believed by the locals to drag fishermen into the depths. It may not be a man eater but at lengths of 3.6 metres it is certainly a formidable hunter, if only of smaller fish. In Brazil another species of oversized catfish which can also be confusingly called the goliath catfish, the piraiba, has been known to eat not just fish but at least one unfortunate monkey- although whether the monkey was already dead or instead was actively killed by the piraiba was unclear. The piraiba's main diet is less exciting though, other fish. That catfish commonly reaches two metres.

Large catfish exist in rivers throughout the world and one of the least explored of these areas is one of the longest, the Mekong River- running from southern China, through Laos, Thailand, Cambodia, all the way to Vietnam. In 2005 a giant was caught in Thailand. It was a species that was well known to the local people; it can even be found in cave paintings three thousand years old. At 2.7 metres long and just under three hundred kilograms the giant catfish swimming in these waters competes with the Wels for the title of largest catfish. It may not be as long as the Wels but it is significantly heavier. Official record keeping only commenced in Thailand in 1981 and the fact that the river extends across so many countries where records of this fish have not been kept for as long suggests that it could potentially grow much larger, particularly in earlier decades before fishing pressure was as intense. Yet, despite first appearances, the giant Mekong catfish is a toothless vegetarian that couldn't eat a duckling if it had to. Its head is misshapen and at first sight looks like it has been attached to the body upside down;

while its eyes are low and only just above the jaws. This is not its most unusual feature. It has no barbells. It's a catfish without the catlike 'whiskers' for which the group is named for. So what makes a catfish?

Catfish are actually just one group of four closely orders. This consists of characins like the piranha, cyprinids like the goldfish, catfish, and the smallest of the orders, the milkfish. Together catfish and their relatives are the most successful of the freshwater fish. But like the Mekong catfish not all members of the family are distinguished by barbells. To complicate matters other completely unrelated fish like the sturgeons also have barbells. What really unites the group is the unique structure of their ear. Ears are not something naturally associated with fish. That is because water carries sound so well that they have no need for external ear drums. But some fish have nonetheless found ways to greatly improve their hearing.

Sound waves vibrate against not just a fish's head but also upon its swim bladder. Normally this noise is just lost but in these fish a complex array of bones in their skull redirects the sound waves hitting the swim bladder to the brain; making their hearing exceedingly accurate. Catfish, characins and cyprinids all have this arrangement of the ear, what is called a Weberian structure. This Weberian structure basically replicates the design of our own ear drums, with its hammer and stirrup structure. Carp in particular are renowned for their precise hearing. But milkfish, the most distant of the four families, aren't. Their ears are connected by just a bare framework of bone, the most primitive arrangement, and the milkfish do indeed seem to most closely resemble the common ancestor of the group. There is indeed some genetic similarities between them and herring- their very distant relatives. The surviving members of this group give us some of the closest insights available into the lives, and behaviour, of some of the first catfish.

Adult milkfish patrol the tropical seas of much of the Indo-Pacific in large shoals, sucking up plankton and dead floating matter. It is a way of life that they have perfected to such extent that there are only minimal differences evident between these gentle filter feeders and the fossils of the first of their kind, some hundred million years ago. These large silver fish move through the oceans with such ease that it is hard to see a visual resemblance to the slovenly bottom feeders that make up the majority of the catfish. Juvenile milkfish often come into mangroves however and it is there they begin to demonstrate how, perhaps, the ancestor of the first catfish entered the fresh waters and took up its distinctive lifestyle. The trees that make up the mangroves hold masses of turbid mud within their roots which quickly become overgrown with algae. This green flourish is incessantly devoured by the milkfish but there is more nutritious food hiding between the roots themselves. Boring through the mud are hordes of small invertebrates, creatures which are much easier to digest than the tough fibrous algae. To reach this new food source the juvenile milkfish have to push their heads into the mud, shuffling away till they have attained their meal.

Beaked Sandfish, South of Fremantle, Western Australia- *J. Martin Crossley*

A close relative, the beaked sandfish, spends even more of its time on the bottom. Species of beaked sandfish are found near the coast throughout the world; in Africa one of the better understood species, popularly called the mousefish, feeds almost entirely in the substrate. The change of diet has led to the mouth moving back so that it is underslung like a shark, making it easier to snatch up passing prey. Wriggling inside its mouth is a barbell. In this protected position its of little use for helping a catfish find its way around or even electrical detection, such as a shark uses. Instead it's covered with taste buds and allows the fish to easily taste for food, acting like a second tongue. But this barbell is only effective for food that is already inside the mousefish's mouth so in front of its nostrils is another barbell that helps the fish to taste and identify potential food hidden in the mud.

The milkfish family is unusual in that most of its members stay in the ocean their whole life. All of the other extended catfish group, with the exception of two small families, are restricted to freshwater. The red tailed catfish is a typical example of the group that lives throughout the northern countries of South America. It can be seen patrolling trees that overhang its rivers but it is not interested in any insects or spiders they may harbour. Instead it is the fruit it is waiting for, which drops with a large plop into the water. The red tail has the largest mouth per body size of any fish and easily crushes the fruit into pulp. Like the mousefish though this catfish comes to the surface only occasionally and it obtains most of its food from the smaller fish and worms that bury themselves in the mud below. Its superior hearing allows it to hone down prey but it has another technique. Extending around its mouth is not one but six long barbells. These act a bit like an ant's antenna, flicking against obstacles in the darkness, but the taste buds covering them allow them to determine which obstacles may prove edible. They continue wriggling in the substrate until they finally locate prey, which the red tailed catfish can easily seize with one snatch of its mouth. Barbells are instrumental in the survival of most catfish and nowhere has this been better studied than in the species of North America.

Unlike the rich waters to their south the entire continent of North America has just a single family of freshwater catfish, the bullheads. Like their southern cousin barbells are crucial for the brown bullheads to navigate and feed in the darkness but they also have other advantages. No true catfish have scales; instead most species are covered in a smooth slime coated skin, similar to that of the hagfish. What is found in the bare skin is much more interesting. Barely visible running from the face across the flanks is a row of peculiarly shaped cells that stick out like polyps. Scientists engaged in studying these fish were at first puzzled as to the nature of these cells until they were discovered later in a completely unrelated animal; humans. A network of the polyp like cells was found in the tongue, only in people each is protected in little capsules. The strange cells shared by both people and catfish were tastebuds, only in the fish they spread across most of its body.

Like all fish, and indeed all vertebrates, catfish still have taste buds in their mouth as well as outside their body. The taste buds are variable in sensitivity depending on where they're located. Those around the body and barbells are receptive mainly to sweet sensations, providing the first clue to a potential meal. But in the mud clogged depths much of this matter could be either decayed or soiled. At any rate it would be hard to distinguish the taste of potential food without the influence of the stagnated water. However the back of the mouth is lined with taste buds which are sensitive, and highly reactive, to any sour flavours. Anything swallowed that is either inedible or decaying is quickly tasted and spat out again. Catfish have the most sensitive palette of any known animal, crucial for distinguishing food from waste in polluted waters. They might be able to find food easily but they are also at risk of being eaten themselves.

One species of bullhead, the North American madtom, has armed itself with a solid spine in front of the dorsal fin and an extra spine in front of each pectoral fin. Madtoms' spines have an additional defence; around each spine are venom glands. Only the thin scaleless skin keeps the venom entrapped in the glands. As soon as the spine is touched the skin breaks and venom seeps out into the culprit. The Bullhead family which the madtom belongs to is successful but limited to North America. Other continents have been colonised by catfish from different families with a completely different range of behaviours.

Sharptooth catfish are key predators across most of southern Africa. In Botswana most of these catfish face a particular challenge. The country's main wetland is a seasonal swamp called the Okavango. Each year the surrounding Kalahari Desert is inundated as the Okavango is flooded by water flowing down from Angola. Papyrus shoots up across the desert, sheltering shoals of tiny fish and the catfish spread out in turn after their prey. The wet season doesn't last long and soon some of the catfish are cut off in isolated pools. Under the relentless sun the pools continue to shrink and the trapped fish converge. There is a brief boom time as the catfish feast on the other fish caught with them but soon it becomes more serious.

As the weeks drag on the mud begins to dry around the catfish, like cement. If it dries around their gills the catfish will suffocate. The withering grass surrounding the pool begins to ignite. While the catfish are relatively safe in their pool the reduction in shade causes it to evaporate even faster. Ash begins to coat the pool as well as mud. In their desperation some catfish try to leap out of the pond in the attempt to reach a larger pool. But by this time nearly all the large pools have either dried up and those that aren't are isolated from each other. The mass of writhing catfish attracts fish eagles. Even at their weakest the catfish are still not easy meals and many are grabbed by an eagle only to be dropped again. The sharptooth catfish wriggle together and wait until the reprise of the flood comes. When the streams flow into their ponds they will once again slip through the papyrus, reigning again as supreme hunters; indeed so effective are these catfish at surviving challenging environments that they have been introduced to several countries- unfortunately often rapidly proving themselves a rather invasive species, as they already have in India.

Sharptooth Catfish, Jaisalmer, Rajasthan, India- Chitra Shanker

There is another enterprising catfish that came to the notice first of the ancient Egyptians, then later the Romans for the shocking effect it had on them. Like some of its relatives it uses electricity to seek out food but it also uses it in attack. The muscles in its skin are strictly arranged into bulky cells dividing up the electric stimuli in its body. This effectively spits the catfish's body into polar north and polar south segments, just like a battery. Even more like a battery the catfish can release its energy and the water naturally carries the electrical charge through it. Any fish caught in the charge will be electrocuted. Fish are not the only victims to be stung by the fish. Many a fisherman must have pulled in one of these electric catfish only to drop it inadvertently back in to the water, out of sheer shock.

Electrocution is suitable at any time for defence but to be useful for hunting the electric catfish has to pick its timing. During the day most fish can evade the catfish and slip safely between the reeds. Even when it is not hunting the sheer presence of the electric catfish is enough to drive many fish to attempt to leap out of the water into alternate rivers and ponds. Those that do not manage to escape will soon have to contend with their vigilant stalker. At night most fish can rely on their lateral line to warn it of approaching threats nearly as effectively as they use vision during the day. At sunset however, when most fish are alternating between these two senses, the animals are disorientated and it is now that the electric catfish begins its hunt. The electrical stimuli it uses to find and kill prey are equally powerful regardless of time of day so it is not encumbered by the change in sunlight. When it sidles up to its prey a single touch of its skin will release 400 volts into its victim. Powerful as this there are other fish across the Atlantic with an even more deadly control of their electrical energy.

Patrolling the Amazon is a slate coloured fish that has confused and stunned both locals and scientists for many years. To some it looked like a peculiarly long carp, others even thought it was some sort of lungfish. So, not surprisingly, the creature got lumbered with a name that was only half right, the electric eel. While not remotely related to an eel it is certainly electric. The two and a half metre catfish cousin moves like a torpedo above the substrate, seeking for a way to fill its endless hunger.

Unlike the electrical sensory pores in most fish those in electrical eels are highly obvious, appearing as rows of deep pits running over the head and are especially visible around the orange lower jaw. The muscles that produce its electricity are concentrated not in the skin but in the tail and what a tail it is. More than 80% of the fish is made of this extraordinary electrical extension.

The electric eel produces a phenomenal amount of electricity, up to 600 volts. This is easily enough to kill most fish and stun a human. But being a living battery comes at a cost. Any time the eel curls around or wriggles its spine it disturbs the flow of electricity and is unable to detect or stun. In order to use its special weapon it cannot swim like other fish, flexing its body back and forward. Instead, like the electrical aba, it uses a specialised fin to propel it forward without moving the rest of the body. In the aba this fin runs across the spine, but the electric eel, having developed such a feature independently, has the fin located below its abdomen. The fin stretches from behind its head down to its tail. By continually rippling this fin it moves leisurely through the water, almost as effectively as other fish use their pectoral fins and tail. By restricting its swimming to one fin it avoids moving any of the muscles that would impact the use of electricity. One unavoidable consequence is that it moves slowly, something that for other hunters would result in starvation. The Torpedo ray relies on its turn of speed to snatch fish inside its fins and electrocute them. Even the electric catfish has to touch its prey in order to stun it. But the electric catfish does not have the charge of four electrical power sockets.

The electric eel slides up behind a moderate sized fish. Most fish, and indeed basically all prey animals, have a flight zone around them that is based around their own size and speed. Any potential threat that enters the flight zone triggers the fish to flee but hunters that are outside the flight zone are basically ignored, as they are typically too far away to be of risk, yet. But by zapping an electric eel can begin its attack earlier than its quarry expects and thus take them off guard.

The constant amount of electricity running through each animal has another purpose. It allows the electric eel to track each other, like the radar in a submarine. Finding one another is essential for breeding but *that* has its own complications. Electric eels face a problem shared by few other fish. Because of their large size and unusual body shape their gills alone are not enough to support them so, like the saratoga and pirarucu, to avoid suffocation they have to breathe air. Their developing young in the egg have the same respiratory difficulties only they can't head to the surface to breathe extra oxygen. Poor buoyancy means that the eggs naturally sink to the depths of the river. In order for their young to survive their father takes an innovative approach. Before courtship he sucks in air and blows bubbles underwater creating a cylinder nest in the process. Thickened saliva holds the construction in place and as the male continues releasing streams of bubbles it begins to float. When the larger female arrives her three thousand eggs or so will be ideally positioned to receive enough oxygen while they develop. This is about as social as these oversized fish will get.

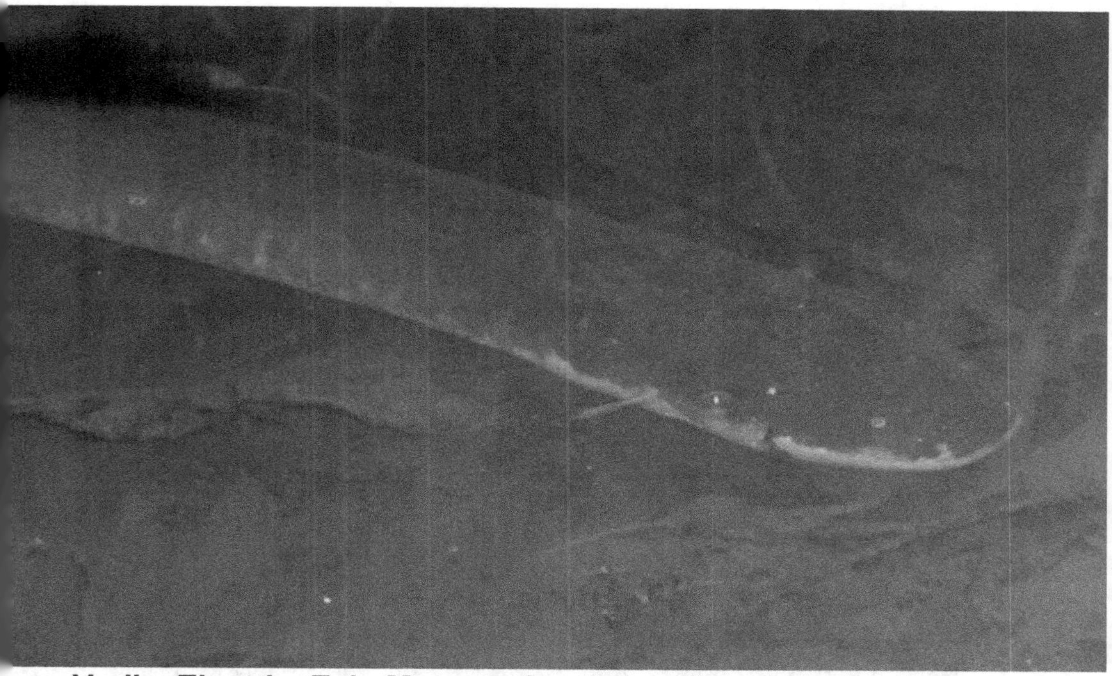

Vari's Electric Eel, Morona Santiago, Ecuador- *Brothers Mathieu and David Sannier*

Being so large electric eels are primarily solitary animals and must maintain vast territories to feed in. As soon as they feel the strong vibrations that indicate the presence of another of its species the eel will either change its course or fight for that stretch of territory. But the electric eel has close relatives that use electricity in far more tactile ways.

The electric eel is the largest member of an unusual order of family closely related to the catfish; the knifefish. Other knifefish don't have the powerful muscles to release jolts of electricity into the water but they are highly sensitive to other electrical sensations. As they glide over the ground like metal detectors they pick up the movements of any small animals. In the depths of the Amazon this skill has been put to an additional use. The continuous release of tannins from rotting trees submerged in the river blots out the light. With their electrical sensitivity Glass knifefish can find not just their food but each other in the darkness. The glass knifefish forms into tight shoals. As many as a dozen or so different individuals flit in and out of the flexible groups. They continually release little beats of electricity to stay together. Not all knifefish are equal and some assert their authority with a stronger current of electricity. In order to avoid interference with each other's signals the knifefish typically vary the frequency of their beats.

All the knifefish in the shoal change their electrical frequency when they come into contact with one particularly large individual, the group's dominant male. He may reach 35 cm, about ten centimetres longer than most of the females. His sheer size allows him to produce electrical pulses which are strong enough to deter most young males trying to challenge his authority. But young males grow and eventually one will become old enough that his superior can no longer intimidate him by size alone and he refuses to change his frequency. The two males commence sparring by sending powerful beats of electricity towards each other until one subsides. If neither retreats they may turn to physical violence and grip each other by the mouth until one, utterly exhausted, finally gives up. In this way the Glass knifefish manage to maintain a society in the darkness based entirely around an undercurrent of electricity.

The Black Phantom Knifefish by contrast is solitary and much larger, about half a metre long. Being larger in turn makes it harder to catch enough food to fill themselves. Even with the advantage of electrical tracking their sheer size acts against them as they squeeze through the debris. Like the electric eel their hunting success has had to come with the sacrifice of maintaining a society. The black phantom is a dedicated loner, only coming together briefly to breed. Although large for a knifefish it may still be attacked. It has an unusual defence; most of its vital organs are compressed at the front of its body hence it can withstand having more than half of its body bitten off by predators. Not only can the knifefish survive such an assault its body has impressive repair abilities. Soon after the attack the knifefish can regrow its tail entirely, leaving barely a scar behind as part of the process.

Unlike the knifefish, not all catfish, even in the Amazon, are sensory hunters. Some, like their relatives in the Mekong, have become strict vegetarians. They have formed one of the most populous groups of catfish, the plecos or Loricariids. Plecos are also called bristlenose catfish after the obvious bristles sprouting out over their face. In the males of some species they resemble the antlers of deer, and are displayed aggressively before males come to blows. There are so many species that are yet to be identified by science that those in the aquarium trade have developed their own peculiar way to distinguish between types. Each different form of pleco is given its own number with an L in front of it (L for Loricariid.) Currently there are six hundred different plecos recognised in this manner. It is likely that many of these will prove to be only different subspecies or regional colour variations but it does go some way in displaying the sheer variety of form in this group.

Plecos, for the most part, share one highly modified sense organ that extends well beyond the limits of any other catfish. Electrical sensitivity is of little use to these fish as the cellular bodies of plants do not use electricity nor do plecos need to sneak up on their meal in the dark of the night. They do need to recognise predators however and distinguish which submerged plants are palatable and which are poisonous. To do this they have eyes like no other vertebrate on the planet.

Bristlenose Pleco, aquarium Brisbane, Australia- *Brothers Timothy Farlow and Nicholas Kable*

Most other backboned animals can control the amount of light that enters their eyes by reflexing their pupils. By this process the pupil can reduce glare from the midday sun or alternatively allow the maximum amount of light in during the darkest nights. Plecos however have no pupil reflex so they have an alternate way to protect their eyes. Protruding over the iris of the eye is a large colourful lobe. When hiding in the depths under the mud this lobe is barely visible but as it enters more transparent water the lobe undergoes a substantial transformation. It begins to swell so that it covers most of the eye, leaving a small enough gap above the iris so that the pleco is not blinded. In this way the pleco can travel from clear to completely turbid waters throughout the day and still be able to distinguish the edible plants surrounding it. The pleco tends to shun the most indigestible plants but even so it still digests some toxins through its diet. In order to survive it has to find a way of neutralising its meal and it has one. Also submerged by the tropical floods are rotting logs which contain minerals that suitable for counteracting the poisons of living plants. By grazing on the bogwood the plecos are able to continue foraging without ill effect.

One of the largest plecos, does not just chew on bogwood occasionally. It is its regular diet. The giant panaque is one of the most peculiar of all plecos, it has become to its family what the giant panda is to the bears, a peculiar specialist, both in diet and appearance. No other backboned animal is able to live just on dead wood; even beavers require fresh timber to survive. Indeed, there is debate as to how much energy this fish is actually able to extract from the wood as its gut is the same as in other related species that need to feed on living plants to survive, and there is further contention that it loses weight while feeding on the submerged timber. They certainly do eat algae as well. Other researchers have identified gut bacteria that possibly does break down the wood enough that the fish is able to digest it. What is clear is that the giant panaque does eat wood, its waste has been compared to sawdust, even if we are unsure yet exactly how important this part of its diet is. Consequently the giant panaque has a massive amount of food at its disposal with minimal competition and even that is only from smaller plecos which can easily be intimidated.

Living hidden in the mud causes its own problems. There is a lack of oxygen here so plecos and other bottom living catfish have to breathe air at the surface to avoid suffocation. On the return to the substrate giant panaques are likely to come down in a slightly different

Whiptail Catfish camouflaged, and in Close-up, Tahuayo River, Peru- ***Xaime Beiro- Fishipedia***

area and encounter others of their species. Panaques being so large are highly territorial. Their aggressiveness is further reinforced by formidable spines that jut out of their gills and pectoral fins. These weapons are efficient defence against predators but are just as effective when turned on each other. It is little wonder that with these colossal giants around that other catfish have found ways to be inconspicuous.

Whiptail catfish are experts at imitating twigs. Their snout protrudes like a miniature branch in front of their slim bodies and even their tail has a pair of thin projections that flick out behind it. The whiptail looks so similar to a twig in fact that it is often only when it darts away that it finally reveals its presence. The female bristlenose pleco camouflages nearly as efficiently, in this case with bogwood that the giant panaque grazes on. The male bristlenose is rather more obvious due to the adornment of a clump of soft projections that stick out like antlers from his head. This facial decoration stands out to predators but it also displays his dominance over the territory to other males, and he will aggressively rip into any would-be intruder. As an extra defence the bristlenose has spines like the panaques that it can inflate to deter any predators. The facial protrusions also allow females to easily identify the fittest mate.

Another catfish finds protection by staying out of water. Wriggling through the damp leaf litter the tarumazinho catfish initially looks like a red worm. Its head is unmistakeably that of a catfish with its barbells seeking out any hidden prey. Its scarlet body might not resemble typical catfish camouflage but surrounded by shrivelled auburn leaves it blends in surprisingly well. Most of the forest surrounding the Amazon River gets inundated regularly enough to keep the ground damp even during the dry season. But the raising water raises an alarm to the tarumazhino which tries to escape the torrent by travelling to higher ground. For this catfish while water is essential to keep its body moist, swimming is not something that comes naturally for this fish. It also feeds on the land. Growing on the protected underside of rotting leaves are large flourishes of algae which can also survive out of water due to the high humidity levels. At a centimetre long the minute tarumazhino easily slides between the foliage and gorges away on the algae, with little chance of being detected.

A slightly larger catfish uses its numbers to protect itself, moving through the water en masse. But this species' appetite will not be sufficed by either plants or bogwood. Instead it is on the lookout for bodies, both those that have died naturally and the kills of others. It does not take the vulture catfish long to clean up after others, in under an hour a shoal can get through eighty kilos of prey. They will take a chance on more lively prey too. Of particular interest to them are the fishing boats that regularly travel through the Amazon and Oronoco. Any discarded entrails or viscera thrown overboard by fishermen at work gutting their catch will be readily consumed but rapidly filling nets beneath the boats are equally enticing. Their slender build allows the scavengers to squeeze inside and vary their diet a little, and just as easily slip back out once the net starts to leave the water.

There is another catfish that has found a sneakier way of making a living. The candiru is a hunter like most catfish but unusually it targets fish significantly larger than itself. Indeed the candiru is one of the smallest catfish; at just a few centimetres long it dwarfs only the tarumazhino. The candiru detects prey not with electricity but by smelling the waste products leaving the gills of larger fish. Despite its slender shape and size this is no gentle creature, it is the vampire of the catfish world. When it reaches a suitable fish it wedges itself into its target's gills. The host fish may wriggle in pain but the candiru won't be dislodged. It has a spine which it raises, lodging itself firmly against the fish. Even if the injured fish tries to dislodge the parasite by rubbing itself along the bottom of the river all it will achieve is simply driving the candiru deeper into its body. Once secure the candiru takes a bite and begins to suck at the released blood at leisure.

There is a problem in the candiru's seeking abilities however. The main waste product it uses to locate the gills of a fish is urea. Urea also happens to be the chemical found in our urine. With the same drive that usually propels them into the gills of larger fish the candiru plunges headfirst into the urethra of any man careless enough to be urinating within reach. When the victim tries to remove the fish it responds by inflating its spine; lodging itself in tissue far less flexible than gills. There are only two effective solutions, castration or surgery. Or so it was thought. When studied by scientists candiru actively ignore urea, or any similar chemicals, and instead hunt by sight- looking for gills. Indeed, the single peculiar reported case in modern times, when the alleged victim stated that the catfish physically leapt out of the water to get inside his urethra while urinating defies both the catfish's limited swimming ability and the laws of physics. Their parasitic behaviour on the *outside* of the body is better known and they certainly can give a painful bite.

Cuckoo Catfish, Lake Tanganyika, Tanzania- ***Photo by Ad Konings***
Vampire Catfish attached to pink dolphin, Peru- ***Thibaud Aronson***

Other larger relatives of the candiru, the vampire catfish, are a little more ambitious, some of them latching onto the skin of significantly larger animals, even predatory pink dolphins and starting to rip in to the juicy flesh beneath.

When it comes to survival cheating may be applied not just to feeding but also to looking after your own young. In Lake Malawi there are cichlids that go to extreme lengths to look after their offspring. Many mothers keep their babies in their mouth and some may even shepherd them between feeding sites until they become too big to hold. During this time they stop eating, in case they accidentally swallow one of their own young in the process.

Cichlids are not catfish but they do attract the interest of one species, the cuckoo catfish. Much like the bird, cuckoo catfish intend to reduce their investment in parenting as much as possible. While a pair of cichlids is courting the cuckoo catfish draw near. Mother cichlids are so protective that they catch the eggs in the mouth almost as soon as they appear. So vigilant are they that during spawning the mother will snatch up any eggs she sees, including those of the catfish. The baby catfish is working to a strict timetable though. If the cuckoo catfish hatches too long before the other young then the mother won't have begun opening her mouth to allow her offspring to leave and feed. If the catfish hatches late it will have to compete with the baby cichlids for food. Worse, if the baby cichlids have already grown and left the mother's mouth before the cuckoo hatches she may start eating again and swallow the juvenile catfish in the process.

Ideally the cuckoo catfish will hatch just before the baby cichlids. It may eat a few eggs but there will be enough movement from the baby cichlids to encourage the unwilling foster mother to head to feeding grounds. When she opens her jaws the infants are able to swarm out and feed under her watchful eye. At the first sign of danger they all squash back into her mouth. But the greatest danger for her young is now inside their former safe haven. After finishing the eggs the juvenile catfish has begun to gorge on the other infants. The cuckoo catfish have found the ideal way of not only housing their young, but priming their hunting instincts.

Not all catfish of Lake Malawi act this way. There are some catfish that not only guard and care for their young but others of their own species. Occasionally some of these babysitting catfish take on the charges of cichlids surrounding them, including those that are targeted by cuckoo catfish. Perhaps it is this early association with such benevolent catfish that allows the cichlids to be manipulated so easily later in life.

From Venezuela down to Argentina live other catfish with equally unusual breeding behaviour. The bronze corydoras is an armoured catfish that feeds mainly on snails and other secretive invertebrates. Hidden amongst the mud it swirls around it, it is scarcely surprising that its behaviour barely drew the attention of scientists or indeed anyone in particular. That was until its mating was observed in a laboratory. After a brief courtship both catfish came to rest on the ground separately. For most this would be the when the female would release eggs and the male would fertilize it. But in this the corydoras proves anything but typical. The male begins to release his sperm but the female does not respond to this in a way he might want. The female apparently has more of an interest in her appetite than reproducing. She simply drinks up the sperm and swims away. The female continues swimming through her territory as usual with the extra parcel of nutrition inside her. But that is not the end of the story. The sperm passes through her digestive system as an undamaged package sticking to the female's pelvic fins once they are excreted. The corydoras presses her fins closely together to hold the sperm in position. It is only now that she finally releases her eggs. The eggs press into the store of sperm grasped in her fins and are finally fertilized. By her unique trick the female has been able to travel far enough away from the original spawning site that her young can avoid being eaten by their father. The mother leaves the eggs attached to any leaves overhanging into the water or sticks them to large pebbles. In less than a week the young will hatch and begin to hunt out prey. But their mother has already left.

Two families of catfish have left the relative safety of freshwater for the sea. Inland most of the predators large enough to threaten a catfish are land based; most of which can be outmanoeuvred by swimming deeper. In the ocean there are plenty of specialised predators that could easily make a meal of them. So, they have become some of the fastest of all catfish, using their streamlining to full advantage to speed away from harm.

Gafftopsail catfish use their large dorsal fin to stabilise them as they dart at speed after their prey although, despite their popularity, research on their top maximum speed remains lacking. Their solid streamlined build has more in common with other active coastal species than it does more sedentary catfish but the mouth, and those long barbells surrounding it, makes its ancestry immediately obvious. By way of defence the gafftopsail are armoured with venomous spines like the American madtoms, although sharks remain a threat- one that more experienced individuals become more adapt at avoiding.

In the Pacific eeltailed catfish also rely on their speed and spines to withstand attackers. Even this is not enough to deter some predators. An adult eeltailed catfish may have to turn on its attacker and fight for its life. The young are not large enough to defend themselves as effectively so they have taken additional precautions. Unlike other catfish they have had to override their solitary tendencies by gathering in immense shoals. Other shoaling fish feed at midlevel, sucking up microscopic plankton surrounding them. The eeltailed catfish remain bottom feeders like most catfish so to feed while part of a group takes compromise.

Eeltailed catfish- *Beaezor*

As morning comes up the young catfish begin grubbing on the ground en masse. Like a well-run army all the individuals face the same direction and move at the same speed. The catfish are gathered so closely together that they can only snatch food directly in front of them. With their barbells they quickly detect any hidden prey. The catfish are such highly efficient feeders that were they to move forward in the standard fashion those at the middle and back of the group would starve. Instead as soon as they've finished feeding those at the back of the group swim over their companions to a fresh piece of ground. As they land their companions have already begun to swim over their heads, aiming for untouched seafloor just ahead of the group. By this stage the original front runners of the group are now at the back and swimming over the shoal to feed in a new position. In this way the catfish create the illusion of a giant billiard ball rolling out over the seabed.

As they grow the eeltail become as aggressive and solitary as their freshwater relatives. It may seem surprising then that they show strong parental behaviour. Most marine bony fish release their eggs to float amongst the plankton. The eel tail mother does something different though, she leaves the eggs for the male to look after. For a month he guards some fifty eggs inside his mouth, ignoring any prey as he goes through his vigil. Even once they hatch he can't feed yet for the young will stay with him for about three weeks, sheltering in his mouth when danger threatens. The male can only remain social, and hungry, for so long and if the young do not leave after those three weeks he may take a few of them to make up for his month long fast. Curiously enough the unrelated gafftopsail catfish also resort to the same behaviour of fathers mouthbrooding.

Eeltails travel vast distances during their life. But most of that travelling is restricted inside their territory on the reef. There are other catfish that travel much further. In the Amazon, the great homeland of the catfish, one species is preparing for a migration that is longer than any other freshwater species on the planet.

The dorado catfish, a close relative to the giant monkey eating piraiba, begins to assemble in the river's estuary in Brazil and start to swim into Peru. Like the Northern salmon by swimming against the river the catfish face mighty waterfalls and a current that is constantly working against it. But the catfish also faces another obstacle basically unknown to the salmon, altitude. The fish are working their way into the lower range of the Andes, the highest mountain range in the Americas and the longest range in the world. To reach this point they will have swum nearly 3,220 kilometres. Their offspring when they hatch will be washed down to the estuaries in Brazil only to undertake the same migration when they mature.

In Asia are distant relatives of the catfish that live at even higher altitudes. Loaches are slim, cigar shaped fish that cling to the edges of rivers with their powerful mouths. By slowly sucking their way upstream they can head high above the countryside, gently following the rivers back to its source. Such a process is necessarily a long one, particularly when the loaches dare to tackle such obstacles as waterfalls that lie in their path. Over generations this family has penetrated higher into the mountains than any other group of fish. The Tibetan stone loach is the species most adapted to alpine regions and it is also a great traveller. Commencing in India the loach follows the Ganges into the highest mountain range on earth, the Himalayas. After several kilometres the fish is already higher than most of its relatives. Only when the fish reaches the dizzying heights of 5,200 metres above sea level is it finally stopped by water so cold that it refuses to melt. It is a fish truly worthy of the title of highest fish in the world.

Another loach lives in conditions that could scarcely be more different, the sweltering caves of Thailand. The outside humidity is so high that even deep in the caves enough moisture condenses onto the fish that it can survive with just its underbelly in the water. It climbs over rocks and underground waterfalls with its fins unfurled in such a position it has been named the cave angel. Its body is much broader than most loaches, and it takes full advantage of its size by using its larger surface area to detect passing prey. Like many catfish the cave angel relies on its extreme sense of touch to find food in the darkness.

In more typical loaches their slender and elongated shape has the unfortunate side effect of compromising the space available for its internal organs, in particular the swim bladder. For most fish the swim bladder naturally expands and contracts as the fish changes its position in the water. But in loaches their swim bladder is stretched out, like most of the other organs, and takes longer to adjust.

The most extreme restriction occurs in a European species, the weatherfish, and has resulted in rather dramatic behaviour as a result. The delicate organ is positioned in such a way that it is encased in solid layers of bone and consequently unable to adjust at all. This doesn't affect the weatherfish's buoyancy as, like most other loaches, it spends most of its time at the bottom.

Tibetan Stone Loach, Kazakhstan- ***Kudaibergen Amirekul***
Spotted Weatherloach resting- ***CreativeNature_nl***

Curiously it is a relatively simple phenomenon that most impacts the weatherfish's swim bladder; rain. Prior to a heavy downpour the air pressure between the water and the sky changes and consequently the swim bladder begins to swell, placing unbearable pressure on the fish. It has only one way to remove the pressure, driving the excess gas down its respiratory tract. The fish begins to wriggle madly, expelling gas from its anus in the process. Only once the swim bladder shrinks back to a comfortable size does the fish resume regular activity. So accurately is the fish's behaviour attuned to the surrounding weather that they were kept for many years as a rather accurate predictor of incoming rain.

Although they resemble several catfish, loaches actually belong to a separate related group; the carp or cyprinids. Their most distinctive difference from catfish is that cyprinids, including loaches, lack teeth in their jaws. Cyprinids still have some teeth but these are located further down in the pharynx, deep in the throat. Lacking teeth in the jaws prevents most carp from being as aggressive and effective hunters as catfish. Loaches delicately pick up small invertebrates hiding in the mud. This is quite time consuming though so other cyprinids have developed alternative methods of feeding.

Goldfish also feed in the soil but by constantly gulping up mouthfuls of dirt. Goldfish are unusual, even among the carp family, in lacking barbells so they do not know what is hiding beneath the soil until it's within their mouth. Occasionally they snatch up something digestible in the process. But feeding like this leaves the goldfish with a problem, how to spit out the inedible mud and waste without losing its meal in the process.

The goldfish has a rather unusual way of sorting out nutrition from the waste it has swallowed. Nestled at the roof of its mouth is a large balloon shaped muscle called a palatial organ. This muscle acts rather like an upside down tongue, by grasping pieces of food but its great flexibility allows it to select only the choicest pieces. All of this occurs inside the goldfish's bulging mouth in only a few seconds, with just the fish's bulging lips giving any indication of what is occurring. Once it's finished sorting out food from debris it spits out what it doesn't want and resumes gobbling more mouthfuls of mud through its jaws.

Cyprinids are built for speed. This is not so apparent in the goldfish, but then just as with their feeding pattern these are anything but typical carp. The red tailed shark minnows from Thailand on the other hand are streamlined enough that they are named after the completely unrelated marine predators. They gather in shoals below the surface picking off individual insects that dare to venture into the water. Their speed is due in part to living in fast flowing rivers that can quickly wash a potential meal downstream. A pair of barbells lines the mouths of the minnows but they are so short that by the time they detect prey the red tailed shark has already seen it. These skilful little predators are nimble but it is their vivid red fins and black bodies that made them attractive to people, especially those that began to export the fish in vast numbers to the aquarium trade. As a result this once common species is nearly extinct in the wild but exists in vast numbers, possibly in the millions, in the aquarium trade. Nearly the only red tailed shark minnows in their native lands are those that have been bred commercially and reintroduced into the decorative ponds and artificial wetlands of the country.

Barbs are among the most versatile of the group, competing with the goldfish in their omnivorous and adaptable habits. They live in tropical lakes from Africa to Sumatra and can shelter inside caves like their loach cousins. Their diet is equally diverse and they will rapidly consume insects, plants and smaller fish or indeed any other meal they can scavenge. In East Africa they even trail behind hippopotamus, feasting on their faeces and the algae growing between their skin. In an uncanny resemblance to the cleaner wrasse of the coral reef the barbs even enter the mouths of the massive herbivores, scraping away outgrowths from the tongue in order to complete their meal. Unlike other cyprinids the barbs harvest food together in great shoals. Living in large numbers takes a toll, when a small meal is found an individual must either fight for it or miss out. So barbs tend to quarrel; engaging in constant bickering in an attempt to maintain their spot in the hierarchy. The tiger barb of Indonesia is amongst the most ruthless and competitive of all known barb species. Several barbs will turn on a newcomer and harass it with bites so severe that it will eventually succumb to either disease or starve. Even the barbs own offspring is not immune to attack and they frequently cannibalise their own eggs and young. In the wild large shoals prevent the chances of any healthy individual being targeted for long but in captivity small groups of five or so tiger barbs quickly turn on each other until a single barb survives. Other cyprinids are much less blood thirsty.

Bitterlings are small silver cyprinids that occur throughout Europe. In spring they begin to breed in small sheltered ponds. But they are very specific about where they choose to mate. They will only spawn where there are freshwater mussels and only one species is suitable, the swan mussel. Once a pair has found a mussel the female releases a long tube from the end of her body. This tube, an ovipositor, slowly works its way towards the mussel. The mussel remains shut most of the time but it must open periodically to breathe. The female bitterling uses her tube to slowly insert her eggs inside the mussel. She must do so gently though, too much disturbance and the mussel will close prematurely on her precious eggs. The female slowly withdraws and the male releases his sperm in a great cloud, fertilising the eggs in the process.

European Bitterling with its mussel, East Macedonia, Greece- *Riccardo Novaga*

The young bitterling slowly develop inside the mussel. Not only does the hard shelled mussel protect the eggs physically it provides another service. Each time it inhales it sucks a current of fresh air over the eggs, delivering a constant supply of oxygen to them. When the infant bitterlings emerge they are fully prepared for life in the open. It is only now that the swan mussel receives its reward for foster parenting.

When the young fish finally leave their protective nursery they will take stuck with them juvenile mussels that will eventually drop off to grow elsewhere and potentially foster the next brood of bitterlings.

The cyprinids are efficient feeders both below the water's surface and at mid water, places where few catfish venture. Across Europe, Asia and most of North America they are the dominant freshwater fish. In Africa, South and Central America their success is challenged by another group of distant relatives, the characins. These fish resemble the carp but they have two distinguishing features. On their back they have an additional fin, the adipose fin, where they can store excess fat and hence last for longer without a meal. Unlike carp their teeth are in their jaws rather than their throat allowing them to tackle larger and more varied food items. The striped distichodus in Zaire, for instance, are among the most destructive of herbivores, quickly ripping many plants to pieces with their rounded snouts. The African Pike Characin has developed in the opposite extreme as a powerful ambush hunter. It only weighs four kilograms but that is distributed over a seventy centimetre long body and its teeth are sharp enough to rip into any human fingers carelessly dangling into the water. Curiously, it has developed the same form of nursery as an electric eel; by constructing a nest out of bubbles.

The largest variety of characins occurs in tropical America where they have taken on roles that in other areas are carried out by insects and birds. In semi-submerged rainforests of Amazon the annual flooding benefits the fish but leaves the inundated trees with a serious problem. Seeds that would normally drop into firm soil plop instead into the rising rivers. Were the seeds to remain in these conditions they would soon rot and fail to germinate. But few seeds are left unnoticed by the shoals of hungry characins. The splashing fruit sounds like a dinner gong and small fish begin snatching them as soon as they fall. Once the fruit enters the characin's gut it will begin to break down but the seed itself is tough enough that it survives the process. Once the digestion process is finished the seed will be deposited in a small pile of dung full of nutrients to fuel its growth when the dry season returns. Developing fruit suitable for fish means that their seeds have the opportunity to spread across the entire river system and indeed many trees that use this method are widespread.

Not all characins engaged in snacking on fruit are small. The tambaqui patrolling the surface is a similar size and shape to a watermelon, only its colour is closer to a murky grey. Jutting out of its face are two oversized nostrils that lead them directly to some of the larger fruiting trees in the Amazon. Silt in the water overwhelms most scents but the fish's advanced nose allows it to smell a potential meal above the water. They gather around a rubber tree waiting for its nuts to fall. As the heavy nut crashes from the branches the Weberian structure inside the tambaqui's ears allows it to easily detect the sound and pinpoint the food and it acts immediately, barging away any smaller competition. It holds the nut inside its jaws and begins to gradually apply pressure until it cracks the protective shell. As it spits out the inedible husk and swallows the seed, a bulge momentarily rolls down its throat. Up to a kilo of fruit and seeds can be contained within its gut but its superiority in meal processing means that comparatively less seeds will leave its body intact enough to germinate compared to in smaller species.

Tambaqui- ***Diegograndi***

Smaller characins also take advantage of the deluge by feeding on the insects crowding around the fruit trees. Tetras dart from below after any cockroaches or spiders daring to swim between plants. Due to their habit of hunting from below their mouths are pointed skywards just as it is in the plankton feeding milkfish. Indeed most characins feed either at midlevel or near the surface and their diet is quickly revealed by the orientation of their jaws; the higher it is the closer they feed to the surface. Tetras of several different species shoal around each other, each specialised for different diets and incapable of mating successfully. So they advertise their species with their own distinctive body shape, colour and pattern.

Cardinal tetras are among the most recognisable and gregarious of the group, shoaling unmistakeably in their hundreds. They are small, barely four and a half centimetres, but their scarlet underbellies and the blue stripe along their flank quickly allow them to distinguish themselves from other tetra species. Due to their size they are at risk from basically any predator that can reach them. Their only defence is their speed and sheer numbers, consequently many of them are consumed each year. To compensate for this the gaudy tetras also replenish their populations in vast numbers.

Cardinal tetra shoals spawn simultaneously releasing clouds of eggs and milt which mix and drop to the bottom of the river. Most characins attach their eggs carefully to the dense foliage of aquatic plants but in the black-water where cardinal tetras naturally live there isn't enough sunlight for such plants to grow. Consequently the eggs have to make do by clinging on to any submerged pebbles they wash against. The cardinals have been laying their eggs in these dark places for so long that if they spawn in clearer water, such as in an aquarium, the developing young are vulnerable to light rays present in sunlight, such as ultraviolet, and are born deformed or never hatch at all.

Abandoning their eggs in the darkness is a risky strategy for the cardinals tetra; few of their offspring will hatch and of those that do many will be eaten before they mature. A hungry catfish stumbling upon the eggs could be enough to destroy most of the cardinal tetras hard work. Survival of the species in the long term comes down to the cardinals investing in number of offspring rather than time to look after them. But other tetras prioritise differently; putting their effort into just a few descendants that will have a higher chance of reaching maturity. Indeed some tetras go to extraordinary measures in order to protect their young.

The splash tetras of Guyana leave the water to remove their young from risk of aquatic predators. They choose a leaf hanging just above the surface to deposit their eggs on. The problem is that without water to bring sperm and egg together the chances of successful fertilization is considerably reduced. The parents respond to this challenge by jumping in tandem and wrapping their bodies around the leaf. Once landed the female releases a few eggs and the male fertilises these immediately. The parents drop back into the water to breathe again. They may repeat this sequence of synchronised jumping until up to fifteen bunches of eggs have been fertilised and attached to the leaf.

Once out of water the eggs are at real risk of drying out. The humidity ever present in the rainforest slows the process, but this is not enough. The male splash tetra takes it upon himself to keep his developing young moist. He could jump onto the leaf again, using his damp body to transfer water to his infants but this could damage the vulnerable eggs. Instead he uses his tail to continually flick water at the leaf, splashing it at least twice a minute. For the next three days he attends his young ceaselessly, keeping them damp at all times. Eventually the developing young wriggle out of their eggs and use the slip stream created by their father's constant splashing to slide down the leaf.

Inland from the splash tetras lives a rather more formidable fish. The piranha shares with the great white shark the dubious honour of being Hollywood's most infamous fish. Despite their bloodthirsty reputation they are actually more closely related to the nut cracking Tambaqui than to other predatory characins like wolf fish and pike characins. The little known redhook myleus is fairly typical of the group with its pronounced squat body little different from that of its more aggressive relatives. It is only when the myleus opens its mouth that its real nature can easily be identified. It has a mouthful of blunt powerful teeth ideal for crushing and grinding. A dentition perfect in fact for a vegetarian. Indeed more than half of the fifty or so piranha species actually live off fruit for most of their life, rarely eating anything livelier than a drowning cockroach. Yet it is the atypical hunters that have truly, and rightfully so, captured the public's imagination.

The wimple piranha provides a potential glimpse into how some of the most efficient hunters evolved from a gentle, primarily herbivorous, group. Wimples are small, only about fifteen centimetres long, and act like it by skitting quietly behind any shelter. They feed mainly on insect larvae attached to submerged reeds and other plants, although dragonflies and other insects skimming the water are also easy prey. But these larger insects floundering in the water also attract other predatory fish. Most of these are too big for the wimple to compete with let alone hunt. But it has found another way to make an opportunity from these boisterous invaders. The wimple piranha has a protruding bottom jaw which it can turn on the other hunters; ripping into their bodies with a quick flex of its mouth. The impact doesn't kill its victim but instead leaves the wimple with a meal of scales, a rather more nutritious meal than its conventional insect fare.

During the dry season piranha, regardless of diet, have a high chance of being cut off in small pools. When large groups are trapped any oxygen remaining in the water is quickly depleted but this is little danger to the piranhas. Like the sharptooth catfish back in the deserts of Botswana piranhas have special adaptations to survive the dry. Characins have a fluid filled upper jaw.

Red-Bellied-Piranha shoaling- *Cookelma*
►Close up of Red-Bellied Piranha, Cuxiu Muni, Amazon River, Brazil- *Altrendolmages*

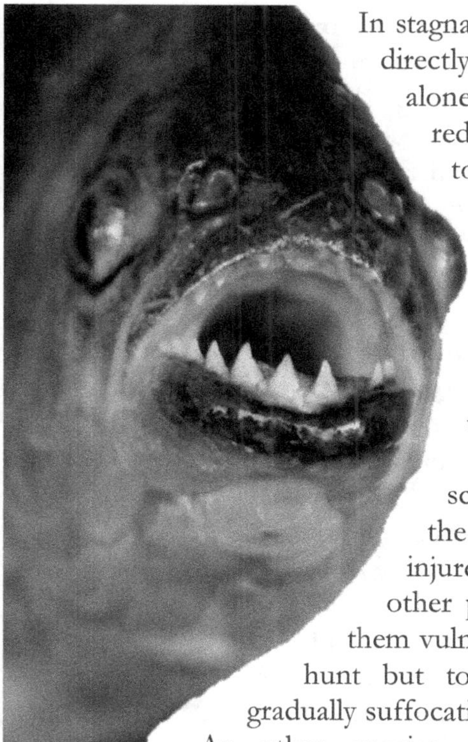

In stagnant water any oxygen that remains is directly below the surface. Relying on gills alone most fish cannot collect this and reduce their activity level as they begin to suffocate. Being able to breathe effectively during these pressing conditions means that the voracious piranhas can turn the drought to their advantage. With their specialised jaws piranha are able to absorb the oxygen in this top layer of water.

Red bellied piranha typically scavenge the kills of larger fish like the river tiger, and finish off any injured stragglers that have escaped other predators. Their small size renders them vulnerable and hence they shoal, not to hunt but to avoid being eaten. A pool of gradually suffocating fish though is a different matter. As other species of fish stagger at the surface, ineffectively trying to get whatever oxygen they can, one piranha will take the initiative to attack, tearing flesh and blood away in the process. Molecules contained in the blood waft into the piranha's nostrils. Ever an opportunist, any piranha nearby will be drawn to the smell. Unlike the wimple piranha these are fish that are not content with a few scales for a meal.

Several piranhas begin ripping at the same prey, yanking it across the foul water. The commotion triggers a feeding frenzy. As the piranha darts forward its swim bladder amplifies a low booming sound being created by its muscles, and the other piranha pick up the signal with their Weberian structure, the same advanced ear structure that first allowed the early catfish and characins to thrive. The result is immediate.

The group converge on the wounded animal, tearing it apart. The victim still struggles as it is eaten alive. The piranha's bulldog like mouth is armed with sharp triangular teeth which slice powerfully through flesh. In the bloodbath feeding becomes a free for all. An accidental bite might as well be lethal. Any bitten piranhas are turned on by their neighbours in seconds. Fierce teeth can't defend you when the whole group turns against you. In less than a minute an injured piranha can be reduced to a skeleton floating to the ground.

It is not only fish that can be killed in these conditions. Capybara, giant members of the guinea pig family that vaguely resemble miniature ponies, may swim across the depleting pools to reach new feeding grounds. Exhausted by the drought they are often anoebic, sickly and covered with deep open cuts. A few drops of blood may leave the animal and enter the nostrils of the voracious fish. Normally the capybara would be safe but under these conditions piranhas converge on the mammal. They can strip the terrified beast to a carcass in less than five minutes. Such fish are also notorious at turning up at the site of sunken vessels and picking off the debilitated victims, although it is unclear how much of that is rumour and how much is fact. There certainly are, sadly, confirmed killing of people by piranhas in this century, although fortunately this remains exceptionally rare.

Fossils reveal an even more terrifying piranha from about ten million years ago. It could possibly get up to 75kg, any piranhas of today's sizes could have been on the menu. And *that* is the uncertain bit. There is a possibility that this supersized piranha was a vegetarian or, even more likely, an omnivore. Some researchers consider that its size and formidable bite would have rendered it capable of biting through turtle shell to get to the flesh beneath. Others contend that it primarily would have targeted nuts and seeds. A clear answer is unlikely to emerge any time soon.

Despite their potential for aggression today piranhas do have a softer side. When the much-awaited wet season returns the red bellied piranha spread out across the Pantanal, Brazil's great wetland. Now they can feed at leisure, picking off insects, browsing lazily on plants, and rejoining the shoal whenever they feel threatened. Those that feel ready to breed will begin to head for sheltered weeds and drowned shrubs at the edge of the flooded forest as their typically bright colour gradually dulls. A pair will begin to circle each other in what was initially considered to be a courtship ritual but in fact it's a territorial behaviour, to guard their planned nesting site. When they are content that their nest site is unthreatened the female will release her eggs. A thousand of these are stuck carefully to the roots of a surrounding tree, but this is not the end of the couple's parental commitment. They have no intention of abandoning their young. They stay with the eggs till they hatch, defending them with their formidable teeth; barely straying from them over the next week and a half. Most other aquatic creatures, wisely, give them a wide berth but a few more plucky fish get close enough to trigger a response from the guarding piranhas. When threatened the parents give a single bite. Unlike the dry season with their young nearby the piranha can't afford for a feeding frenzy to break out. Giving just one bite prevents the hunting craze that would break out if the fish broke into a full attack. Restraint is not something usually associated with piranha but it is crucial for their survival. Piranha are proving themselves the most unlikely and yet among the most considerate of all fish parents.

Chapter 6. Masters of the Reef

Blackblotch FoxFace feeding within the coral- ***Byrdyak***

On land the most productive environments for life are the tropical rainforests straddling the equator. Here high concentrations of sunlight and rainfall combine to create conditions that can support high diversity in both species and sheer number of life forms. In the oceans where lack of water is no issue there is a different restriction on life, nutrients. Most of the nutrients in the ocean have sunken to its depths, well beyond the reach of most marine animals. In chilly temperate regions fierce currents drive these precious nutrients up and fish as a consequence thrive here. Tropical regions are different; here the relatively calm waters are not strong enough to mix with deeper water resulting in vast barren areas; they are the ocean's deserts. Algae near the surface effectively convert sunlight to food but the lack of nutrients restricts their growth. There is an alternate arrangement; primitive animals like sea squirts, giant clams and coral which provide cells that can house the algae. The side effect of the alga being restricted to another organism is that it can only do this where the water is clear and shallow enough for it to photosynthesise; any mud or nutrients washing from the land will kill it. In a few special places a barrier, like a forest, restricts the flow of nutrients into the sea allowing coral to take hold and spread substantially. The resulting reefs compete with jungles for diversity of life. They are an oasis in an otherwise desolate region. The longest reef trails the Red Sea for four thousand kilometres, but it is relatively narrow, barely stretching more than a hundred metres from the shore. The largest reef in the world, the Great Barrier Reef, covers an area of over 225 thousand square kilometres, home to one of the most diverse ecosystems on the planet.

Worms, crustaceans and molluscs all swarm on this reef, as they do on other reefs worldwide. There is great opportunity for any fish that can live here. Catfish are a great success in fresh water but only two families can be found on the reefs. Sharks also cruise the fringes of the reef but they do not feed on either the small invertebrates or the coral. Moray eels are likewise restricted to the few crevices and cracks formed by the reef, grabbing whatever moves close enough as they can't compete in the open. The survival of both groups of hunters is only due to the presence of another group of fish altogether.

The fish that thrive here are bony fish, like the eels and catfish but there is something that sets them apart. At first glance they might look like any other fish but they have one superior difference. Their dorsal fin is not supported by a group of bones but by thin strut like spines, as are its anal fins. It works effectively for defence but it allows the fin to be moved so much smoother, easier and faster than any of their competitors; allowing some of them to become the fastest swimmers on the planet. These are the spiny ray finned fish. Together the group contains about 260 different families. One family alone, the perches, contains more than a thousand species. The Queensland grouper, and its close relatives, are the largest bony fish found on the reef, longer than a man and weighing about 400 kilograms. So large is this grouper that its gills barely provide enough oxygen for it resulting in its rather slovenly nature; shadowing large shoals and occasionally picking off the odd straggler. It could continue doing so for perhaps fifty years if it is lucky. The pygmy goby could scarcely be more different. Not only is it one of the smallest fish on the reef it is among the smallest backboned animals on the planet. Females, which are the larger sex, don't even reach two centimetres. Being so small it lives life at a rapid pace and consequently also has the shortest lifespan of any vertebrate. Less than eight weeks after hatching it will die, of old age. Between these two extremes there is a multitude of spiny ray finned fish making a living on the reef.

Coral, with its multiple arrays of irregular shapes and forms, harbours many invertebrates hiding in crevices that other groups of fish, like eels, simply can't reach. The juvenile bird wrasse can easily squeeze through gaps between the coral to gather up shrimps and other small prey. As it grows this becomes harder for the wrasse's body to achieve but by way of compensation its mouth becomes longer. By the time the bird wrasse is thirty centimetres long its jaws stick out like a beak and can easily be probed between the coral to gather food.

Many butterfly fish pick at the individual polyp animals in the coral. To do this they have precise solid beaks so they can nip through the harder pieces of coral; each species of butterfly fish feeding on a different type of coral. Some specialists exist entirely on this meagre meal forcing them into strictly patrolling their territories. Melon butterfly fish live in pairs which barely tolerate each other as they struggle to survive on their patch of coral. In other species of butterfly fish the mouth is elongated into forceps, providing both a more precise grip and the ability to harvest other prey. The jaws of the long nosed butterfly fish can be a third of the fish's total length allowing it to squeeze between the coral to snatch up hiding invertebrates. Invertebrates are easier to digest than coral polyps and sustain it for longer so consequently this species doesn't require solitary territories and can form shoals. Feeding in groups provides the long nosed butterfly fish with an early warning system; the fleeing of one individual will cause the whole group to flee. All of these fish create excrement and there is even a fish specialised for feeding on this delicacy, the spotted scat that consumes the waste being produced endlessly around it.

There are much larger fish grazing not on the polyps but on the coral itself. Parrotfish bite through the rock skeleton of the coral in their insatiable desire for food. They have massive beaks, allowing them to shear right through it, allowing them to consume the algae that cohabitates with the coral. Deeper inside their mouths are pharyngeal teeth, like the carp, which allow them to continue chewing a meal even after they've swallowed it. There are about eighty different species, each of them, like the butterfly fish, living off a slightly different diet.

The largest and most dramatic species, the bump-headed parrotfish is 1.2 metres long and a vivid purple colour. It is also the most destructive of the group. This formidable species gathers in shoals of more than two dozen and like a herd of bison plough through the foliage beneath them. In a year a single fish can remove a cubic metre of coral from the reef. To avoid completely decimating the landscape around them the herd migrate back to the same hiding place each night only to set off in a different direction the next morning. But the parrotfish also give something back to the reef. As pieces of rock gather in the stomach they are slowly ground against each other and begin to wear away. Gradually they crumble into granules of white sand which the parrotfish excrete in their daily migrations. Over millions of years parrotfish, in addition to the flow of the currents, have contributed to the build-up of sandy beaches across the tropics. But even more significant is their effect on the reef itself. The emergence of parrotfish a little more than twenty million years, along with other large herbivorous fish like surgeonfish, resulted in the expansion of diversity of coral; as much as overgrazing can damage the coral it has an even more noticeable effect on the algae that without the actions of parrotfish would overwhelm the coral. These unassuming feeders have created some of the most spectacular and significant natural wonders of the world.

Bump headed parrotfish are large but they are still at risk from even larger predators. Parrotfish could defend themselves with their sharp teeth but this would involve slowing down enough to face their attackers. Instead they release a great stream of sand from their behinds as they attempt to escape. When this hits a predator's face it acts rather like a skunk's stink bomb, momentarily repelling it. For, in addition to sharks and moray eels, other ray finned fish are hunters as expertly adapted to this role as parrotfish are to theirs; a grouper able to sneak up behind a small parrotfish can snap its jaws around its wriggling body before the herbivore has a chance to make its escape. There are ray finned fish capable of hunting virtually anything that moves. Jacks use speed to cut off their prey's escape routes and vigorously pursue any smaller fish. In temporate seas Sculpins lay camouflaged close to shore waiting for the ideal opportunity for them to lunge forward, scoffing their unfortunate target down head first.

◄Sculpin with freshly caught prey- ***Joe Belanger***
◄Hiding Stargazer fish- ***Joe Belanger***
Dwarf hawkfish, Indonesia- ***TravelSync27***
Bump-headed parrotfish- ***ImageSourceCur***

Stargazers are rather more sedentary hunters and resemble monkfish, even though they are not closely related. Juveniles chase prey in the plankton but adults swell up like balloons and sink to the bottom. They can neither outrun their food nor outwit it so their main hunting feature is their peculiar cage like mouth which can rapidly expand, sucking unexpecting small fish into it in the process. Its eyes pop above the head just as they do in a crocodile, enabling it to overlook its swollen bulldog like jaws and see the right moment to strike.

Back on the reef, Hawkfish cling on to a piece of coral with their specialised pectoral fins and dart forward to snatch up any unwary vertebrates creeping below them. These fish have no swim bladder which allows them to descend quickly but it also forces them to kill their prey quickly and resume their grip on the coral before they are washed away.

It would seem that potential prey would have little defence against such efficient predators. But ray finned fish have as many ways of defending themselves as they do feeding. Many flatfish, like the flounder, can change the colour of their skin to match their surroundings. This works when the fish is simply floating over the sand but for the last couple of centuries, and even more so for the last couple of decades, they have been exposed to something quite different. An increased human population has also meant an increase in artificial waste entering the ocean. Flatfish have superb camouflage skills but modern items like Coca-Cola cans and plastic bags are now creating a rather different background for their skill. Nevertheless, flatfish are proving equally able at survival in a polluted environment; one has even been seen taking on the appearance of a chessboard it was swimming over.

Other reef fish apply more hostile defensive measures. The scorpionfish is so vividly coloured that it blends in with the coral without needing to change its colour. It can simply rest and wait for prey to come within reach. It also has a poison tipped spine that it can spear into an attacker. Unfortunately so well camouflaged is the fish that it may escape the attention of a human walking barefoot over the submerged sand. Only after the fish is stepped on may it actually be noticed. Occasionally this may prove fatal. It is also not particularly effective for an injured fish. Yet there are several fish across the world that hide half buried in the sea bed armed with a poisonous spine ready to ram into an attacker; the lesser weever in Europe, the longspine waspfish around Northern Australia and, most feared of all, the estuarine stonefish that haunts the waters from Australia all the way to China and India. Generally most poisonous fish swim in the open and display their toxins with vivid colours, avoiding both wastage of their toxins and energy from disturbance by unwary predators.

The forktail blenny is yellow with an aqua head that hovers boldly around the reef. It is only small but it does something that most predators simply can't stand. It bites them. Protruding from its mouth are specialised fangs that can be used to inject venom into a victim. This venom is so intense that it can seriously affect a human let alone an animal only the size of a jack or trevally. Not surprisingly other similarly sized fish imitate the forktail and receive the same avoidance by predators as the genuine species. Biting a predator, while effective, still means that the threat has to be close enough to swallow you before you can attack it. An attacking trevally may leave after being bitten but in the process there is still a high chance that the blenny itself will be injured. So other fish have developed ways of poisoning a predator before it even reaches them.

The soapfish has poisonous glands but these are connected to the skin rather than the mouth. Unlike the forktail the soapfish doesn't even have to be touched to defend itself. As soon as a predator comes near enough to spook it the soapfish begins to release the poison rapidly into the water. The poison is not just toxic it is suffocating. Some fish will be so affected that they will be unable to resume normal breathing and will asphyxiate. And the soapfish is not immune to its own poison. Usually it can swim away from its detonation zone but in the confines of an aquarium, or if pinned by predators in a corner among the coral, it can easily fall victim to its own weapon. Poison is also expensive to produce and replace.

The black spotted porcupinefish has adapted an additional solution to reduce its reliance on poison. When relaxed it looks a fairly typical fish, although it would be fair to describe it as a little portly. Pinpricking its bodies are little spines which, although defensive, reduce its streamlining and consequently its speed. It is only when threatened that the porcupinefish truly lives up to its name. By inhaling deep into its stomach its whole body begins to balloon out. Its spines line up evenly around the inflated animal, only the eyes and mouth lack them. The stomach has become so adapted to this additional role that is has lost the ability of digestion and all food is broken down in the intestine, which does not swell. The skin also has a unique two layered design allowing it to stretch smoothly rather than being torn apart during the process of swelling. It doesn't need to release poison into the water to deter predators. Few animals would venture to swallow this prickly mouthful now.

Black spotted Porcupinefish- *Lifeonwhite*

Porcupinefish also catch the eye of people and are often held out of water, carefully, while the human stares in amazement. The stressed porcupinefish begins to inhale but its body is only designed to take in water not air and the effect often proves fatal.

Close relatives of the porcupinefish, pufferfish have a relatively smooth body; in juveniles their spines won't even begin to develop until they are ten days old. Yet this is not the vulnerability you might expect. Like soapfish the puffers defend themselves with a powerful poison but rather than being restricted to the skin it can be spread through its entire body. They still swell up to deter attack but, as a last resort, if the teeth of a predator graze through its smooth skin the poison will almost certainly cause its demise. One puffer, the shosaifugu, has acquired particular notoriety due to the strength of not just its poison but its popularity with the lovers of fine Japanese cuisine. Over time, as stories of the fish's toxicity became widespread in the Western world the fish's name was reduced simply to fugu.

For once the truth about the fugu is as incredible as the stories. The toxin it defends itself with is more deadly than cyanide or even strychnine. The poison is concentrated in the fish's liver but when threatened it seeps out the skin, quickly driving away any marine creatures. Air breathing people are not at risk of inhaling the poison during respiration, only through eating the fish directly. The specialised chefs who take on the serious feat of preparing fugu spend several years just observing before they are permitted to even take a knife to the fish. It will be another couple of years before they will be allowed it to serve it to another human being. Even with the combination of modern technology and years of experience the liver of the fish remains so toxic that it has to be discarded completely; although there are still some customers who deliberately seek out and request the liver due to its associated exceptional flavour, even though it has been illegal for it to be sold in restaurants since 1984. Even with these precautions in place in the ten years from 2008 close to three hundred people became ill after eating fugu and three died, most of these diners having cooked the pufferfish at home. This sorry fact shows clearly the lengths to which a fish will go to defend itself in an ocean filled with hunters.

Other spiny ray finned fish, like the rose snapper, do not delay in trying to engage their predators but instead try to outrun them. But the rose snapper gather in large groups which can rapidly detect both prey and their own predators. The most alarming threat to the snapper is not sharks or barracuda but something much smaller. Among the nutritious plankton being swept through the snapper's mouth and gills is something much nastier. The snapper may first become aware of its presence by a pang in its tongue. It will also start to feel emptier after every meal. Something is stealing its catch. Deeply enfolded under its tongue is an amphipod, the same group of crustaceans that targeted the salps, that feasts both on the snapper's tongue and its food. The amphipod does not stay sheltered beneath the tongue for long. As the increasingly emaciated organ gradually shrinks the aggressive hitch-hiker swells in proportion. For the snapper there is nothing it can do about it. Eventually the rose snapper will lose its tongue altogether by which time the interloper has grown so large, and is so positioned, that it is effectively being used as a replacement tongue- a tongue that occasionally takes its own bite out of the food. This most disturbing of parasites has no intentions of going anywhere.

For all fish parasites are a major concern. Lice crowd into their ears, encrust their eyes and bore inside their skin, making the fish easy prey. These miniature opportunists wedge into any crevice they can find. Lionfish, the free swimming and gaudier relatives of the stonefish, are safe from most predators with their poison glands and long spines. Their delicate fins however have a problem, they are difficult to clean and the poison discourages other creatures from scraping away the parasites. In order to remove this buildup lionfish regularly drag their bodies through the sand and shake wildly, leaving a flurry of dead skin and muck in their wake. This effort can however pay off in more ways than just hygiene. The commotion may draw the attention of small fry eager to investigate, and which are easily impaled with another quick flick of a spine.

Rather than just regard them as a nuisance some enterprising fish have found a way to make invertebrates work for them. Lining the coasts of the Galapagos Islands are rows of barnacles. Being encased within their solid shell is as about as effective a defence as exists. But even so barnacles cannot live for ever and when they perish they leave an attractive home behind for any animal small enough to use it. The withered body initially attracts miniature shrimp and other invertebrates. Their presence in turn attracts a host of small fish that quickly eat the scraps. For most fish this is the end of their investment in the barnacle, but not all. The barnacle blenny fits neatly inside the shell so not only can it eat the last juicy remains of the mollusc but remain there indefinitely. Barnacles quickly colonise suitable areas of the reef so usually individuals of the same age cluster together. This means that the death of one elderly barnacle is regularly followed by that of its neighbour and several empty shells soon lie next to each other. These quickly become filled by blennies, forming colonies with great speed.

The blennies would provide a tasty morsel for any predator that could snatch it but they would have to see it first. Camouflage is its first and most effective defence; its bright colours help it blend, surprisingly, with the algae covered barnacles around it. It only leaves its head protruding from its shelter, snatching up any passing invertebrates. If all else fails and a passing eel decides to try its luck the blenny always can retreat back into its shell, out of the reach of snapping mouths.

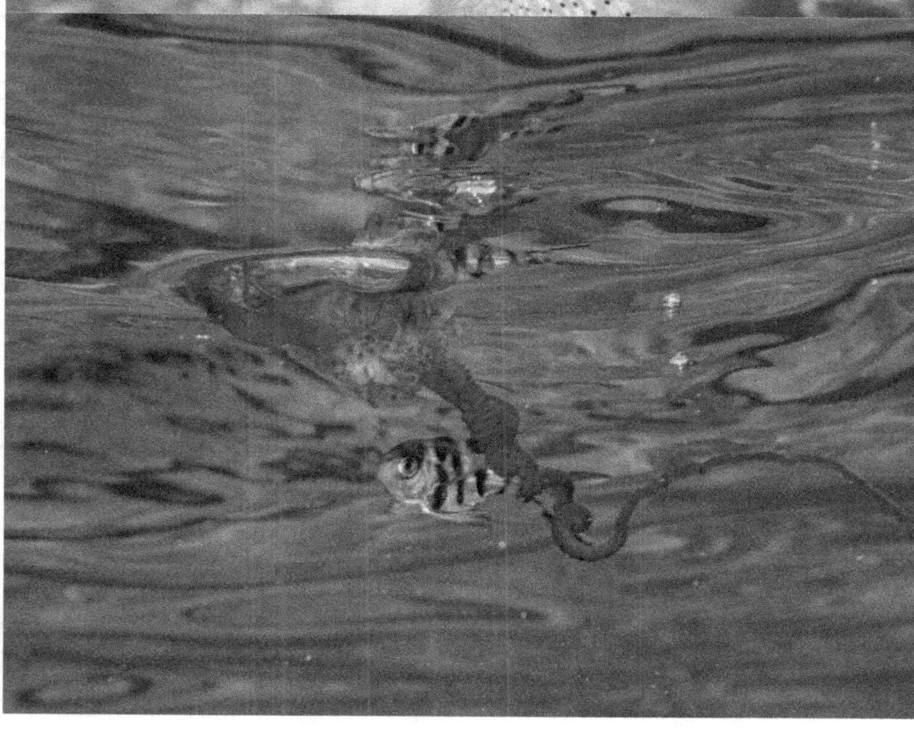

◄Lionfish- ***TravelSync27***
◄Bluebottle Fish, Sydney- ***Henrick M***

Indeed, one of the few times when the Galapagos blenny leaves its barnacle is when a female travels over to the closest male and sticks her eggs to his barnacle. Here the eggs are able to develop nearly as safely as its parent as the father stands guard. But few fish live inside such benevolent accommodation.

Throughout much of the calm tropical oceans there is only enough plankton to feed simple jellyfish. These may appear easy prey but their simple design hides a deadly defence, barbed spines coiled tightly inside their trailing tentacles. A single touch is enough to trigger the spine to spring out into the flesh of whichever animal disturbed it. The spine releases a cocktail of stringent poisons, all designed to break down the immune system of their victim. Among the most venomous of all jellyfish are the Portuguese man of war. Its defences are powerful enough to kill a human. Caught inside its clutches are the bodies of any fish inquisitive enough to venture too near. In this desolate environment such a hoard of food is too tempting to ignore and continues to draw other hapless victims into the tentacles of the man of war.

But there is one fish the invertebrate does not intend on attracting. Just as widespread as the deadly vagrant is a striped blue fish that associates so strongly with the jellyfish that it is commonly called the bluebottle fish. The little fish has enough immunity to the jellyfish to use it as a shelter, hovering precariously between the tentacles. For there are plenty of fish, like juvenile trevally, that could make a quick meal out of a bluebottle fish. In the hunter's haste to ambush the fish it may blunder into the jellyfish before it even notices it. It's the last mistake the trevally will make. The muscular fish goes limb and the jellyfish begins to feed. So does the bluebottle fish, ripping strips of flesh off prey it would stand no chance of catching on its own. But this is not the only way in which the fish takes advantage of its host. As the jellyfish digests its meal it puts the additional nutrients into growing longer rows of dangling tentacles, which allows it in turn to catch more prey. The jellyfish may be immune from larger hunters like trevally and barracuda but it is vulnerable to, of all creatures, the little bluebottle fish that it has sheltered for so long. The miniature fish darts in and around its host, nipping off mouthfuls of the tentacles, swallowing greedily. Somehow it, among all the vertebrates in the world, is able to break down the toxins within the man of war's tentacles. Luckily for the treacherous lodger the Portuguese man of war is able to repair itself as quickly as it is eaten, maintaining a continuous shelter above its head.

The blue streak cleaner wrasse also survives on a diet of invertebrates. These fish patrol the Great Barrier Reef in families, defending territories by flashing their bright blue colours. The colours are also meaningful to animals much larger than the wrasse. Parrotfish and other lumbering fish may soon arrive, stretching out their fins and lying in front of the cleaners. The wrasse takes the cue and starts to wriggle across its customer's body pecking off any food it can find. As other clients arrive they are forced to cue up, all waiting for the services of their daily clean. All this commotion however is not likely to go unnoticed. Sharks begin to arrive, then giant Queensland groupers. Slovenly as the groupers are, a wrasse would still prove a rather easy meal. The cleaners react immediately, heading straight for their most dangerous customers; they can afford to return to the parrotfish later. It must be said though that some of the complaints that most hassle their clients, like embedded fishing hooks and lures, are well beyond the limits of even these diligent creatures.

Essential to the success of these wrasse are their unusual swimming style. Most fish move by flicking either their body or tail from side to side, driving them forward in what is effectively a highly refined wriggling motion. But wrasse and a few of their relatives use their pectoral fins as oars to precisely manoeuvre their bodies. This allows them, for instance, to squeeze through anemone fringed crevices without touching the poisonous barbs.

In cleaner wrasse this swimming technique safeguards them from another peril. Most of the parasites that the wrasse removes cling not to the client's skin but inside of its mouth. Entering the jaws requires complete control of movement. A mistimed fin flick could knock against a gill and trigger the larger fish to inadvertently swallow the wrasse. The wrasse's ability allows them to rotate around the teeth and delicate gills without triggering any adverse response.

The stress of such encounters can have its effect on the wrasse and they may accidentally bite the customer instead of the parasite. Their response is critical. Mature wrasse will stroke their clients gently with their fins, applying deep pressure much as a masseur does in an attempt to appease them. Juveniles lacking experience however jolt and try to anxiously continue on without pause. Annoyed by this poor service the

Cleaner wrasse attending to a potato bass, Durban, South Africa- ***Imagesourcecurated***
False Cleaner Fish, Manly, Sydney, Australia- ***Michal Biniek***

larger fish will leave and seek out other cleaners. If this happens too frequently eventually the cleaner wrasse will have no customers and will starve. In response its competitors will expand their territory and take over the grooming needs of the disgruntled customers; the larger males fighting when they need to take over the valuable location.

The wrasse are not short of competition, even outside their species. Other types of wrasse, shrimp and even seagulls all compete for clients. The sabre tooth blenny looks so similar that many fish easily mistake it for a cleaner wrasse. There are two substantial differences however. Hovering between the blenny's eyes like antennas are two additional sensory organs called cirri. The cirri are thought to assist not taste or electrical sensitivity, as might be assumed, but locomotion, via picking up changes in the water current. The majority of blennies have cirri; each species having a slightly different shape or configuration. The second major difference from the cleaner wrasse is the two fangs protruding from the sabre tooth blenny's mouth. The blenny imitates the dance of the wrasse but adds a critical twist to the cleaning itself. Rather than nip off a few parasites it rips a slice out of its client before fleeing. This is no occasional meal; it is the only way the sabre tooth blenny feeds. The cleaner wrasse that it is impersonating will have to deal with the fall-out, and the possible risk of aggression from the significantly larger customer the next time they encounter each other.

The wrasse doesn't just have to worry about impersonators having a sneaky feed on its customers but also members of its own clan. Flukes and hitch-hiking crustaceans may be numerous but are not particularly nourishing. The most nutritious part of a client is not its parasites but its flesh. This the cleaners receive in small portions from blood filled parasites or the odd lesion but occasionally they decide to collect it directly from its source. Females have a particular need, having high protein requirements in order to produce healthy eggs. The males however will punish and attack any biting females they see so the expectant mothers are forced to fulfil their needs in secret. The ideal opportunity comes during a mass influx of clients. Then while the males are dispersed the females can sneak about and take a few bites before disappearing. Conversely however the males will take bites out of clients without punishing each other; it seems they are accepting of risk taking only among those members which defend the clan. Still, they ensure that after each bite they reassure the client with a gentle fin stroke as if it was just a harmless mistake.

The aggressive behaviour of the male blue streak cleaner towards the females appears even more peculiar when you consider what happens when one of the patriarchs of the group dies. Cleaner wrasse are not born male; the only way that the group's status quo can be maintained is if one of the females changes sex. It appears that the aggression of the males upon the females is the only impediment on them automatically changing sex as they age. With the group patriarch removed the lead female quickly transforms and takes control over the group. Changing gender from female to male is common among wrasse, hawkfish and even among some parrotfish. Although there is a great level of diversity in reproductive strategies amongst ray finned fish most of these can be slotted into two general approaches; to either spawn in massive groups or for fish to court each other and mate individually. Off the coasts of the South Pacific can be seen one of the more dramatic examples of the first approach.

Nassau groupers are about two thirds of the size of their Queensland cousins and well within hunting range of a grey shark. Normally they can back up against the coral and drive away a solitary shark but in the open areas just beyond the reef, where they breed, they are vulnerable to attack from all sides. Consequently, the large green dappled fish gather in massive groups where they at least have protection through sheer numbers. The males circle each other in an uncertain truce waiting for the signal from the females. Releasing her eggs does not come easily so she requires extra assistance to expel them. As she darts upwards her expanding swim bladder increases pressure on her ovaries and eggs begin to stream out of her. The males chase after her and release a cloud of milt as they approach. The females have almost certainty that some of their eggs will be fertilized and develop. But the males have no such guarantee; only those that can keep up with the accelerating females have a high chance of leaving offspring behind. Yet, for all of the groupers, giving rise to the next generation still relies heavily on the chance mixing of eggs and sperm in the current.

Other fish invest more heavily in their young and that means that, the males especially, have to fight for the right to reproduce. The harlequin tuskfish is a wrasse distinguished by the orange stripes marking its white body. Such an attractive appearance is marred by the presence of two jutting canines leering menacingly above the gums. When two males of an equal size come across each other they gape aggressively. They will even turn on their own reflection in a mirror, sometimes with such passion that they smash it. In Malaysian estuaries a rather more gentle looking species, the halfbeak, has a snout shaped like tweezers which it uses to pluck insects off overhanging leaves without being seen. But when two males compete they use their feeding tools just as effectively as weapons. They grip each other around the back of the mouth and begin to twist, applying increasing pressure. As the beaks of the males flex one way and then the other they begin to reach their limit. If he is lucky the weaker male may be able to withdraw but more often the hold on him is so intense that he cannot escape. His beak finally snaps and the victor releases his grip. The victim is not only defeated but will be unlikely to ever feed again. The sole victor of this battle can be confident that all the eggs he sires with the female will be his.

Mandarin fish, Indonesia- *TravelSync27*

Mandarin fish are a combination of such intense colours and patterns that they appear almost psychedelic. They are protected by poison and their appearance quickly conveys this to any curious predators. But it's less effective with each other. The males have unusual eyes that can be rotated in their sockets, like chameleons, which pick up the distinctive movement of any other mandarin fish moving close to their territory. If it is a female the male will commence courtship, if not battle. Both competing males exercise control while fighting, as the accidental release of poison from their skin would severely affect both of them. As they hold each other with their flexible fins and grapple they ensure their teeth do not break their opponent's skin. As a result both finish their aquatic wrestle exhausted but unharmed. Curiously when a female comes near a male engages in precisely the same show of posturing and moving his fins, just at a slower and less intimidating pace

The Siamese fighting fish is so violent that it has been specifically bred not just for its fine colours but for its fighting ability. Over several centuries the aggression of domestic fighting fish has increased to the extent that they will readily tussle to the death but even the wild fish is highly competitive. Again, it is only the mature males that are engaged in such conflict; in both aquariums and their native streams in Thailand the dull coloured females and juveniles live in large shoals. Lording over them is a single blue male with fins projecting from him like great streamers. If his colour and posture wasn't enough to secure his leadership over the group he is also the biggest member. But juveniles grow and eventually one will begin to approach the dominant male in size and colour. The two will start squabbling over the best food and hiding places. If the young male attempts to mate things rapidly escalate. The two males begin circling each other, flushing their slender fins with blood in an elaborate attempt at intimidation. Amidst this standoff one finally snaps and begins to tear into the other. With their short mouths they grapple at their opponent's most vulnerable parts, their fragile fins. Scales and blood begin to rip from the pair as they engage. Unlike the halfbeaks the battle does not end with the complete incapacitation of one male but the weaker of the two is likely to succumb later due to a combination of his injuries and infection.

Competitive as they are the battles of the Siamese fighters and mandarin fish are both overwhelmed by a fish so large it is capable of injuring even people. Titan triggerfish have among the most violent of confrontations. Each testosterone charged male maintains a cone shaped nest, which he constructs with his mouth. He defends not just the nest itself but a similarly cone shaped area surrounding it. Any intruders will be barraged away with such force that an adult human can be knocked unconscious during the process. The triggerfish's position atop the food chain means that he has consumed dangerously high level of certain chemicals, allowing him to inflict potential ciguatera poisoning upon his target with a single bite.

It is important for the male titan triggerfish to be able to easily recognise females so that he does not end up attacking them by mistake. Like nearly all bony fish he has excellent colour vision and is able to instantly identify a female by her subdued brown colour. A considerably smaller fish, the boxfish, also uses colour to distinguish the sexes, in a Caribbean species the male is golden while the female is a delicate shade of lilac. Like mandarin fish boxfish are protected by poison glands in their skin so they are safe to pursue courtship in the open. For extra protection their body is also enclosed in a rigid skeleton which restricts their movement to a slow paddling pace. Undeterred by this boxfish pair off confidently, trailing each other in what appears like a curious version of 'follow the leader' or the waltz until they decide it's the right moment. After spawning they will separate just as decisively and return to seeking out invertebrates in the sand.

Grey angelfish also commence mating with an initial courtship where the pair closely follows each-other's movements. Unlike the boxfish they are monogamous, remaining faithful to each other throughout their lives. In order to do this, they fiercely maintain their territory; the male targeting passing males while the female equally effectively evicts members of her own sex. Only by such unsociability with the rest of their kind are the angelfish able to maintain their pair bond. Colour is used to signify not just gender but also readiness to breed. In some marine angelfish the parents and young look so different that they were originally considered different species. This visual difference prevents the fish from needlessly attacking young too immature to reproduce.

Hamlet fish have found a solution of sorts to this competitive behaviour. The animals are simultaneous hermaphrodites, with both male and female sex organs occurring and active in the same individual. While this reduces competition between genders the relationship is still not entirely conflict free. Eggs take more energy to produce than sperm and both fish are initially reluctant to release any. To test their compatibility and level of trust they perform an elaborate courtship dance, twirling around rhythmically. Eventually one of the fish releases a small stream of eggs into the water. Encouraged the second fish fertilizes them and releases some of its own eggs. Again, only a handful of eggs are produced each time. This process will continue over several nights. It ensures that both parents make an equal contribution to their offspring but it is also pedantic and time consuming. It is scarcely surprising that no other group of fish has adapted it.

Mating Yellowbelly Hamlet, Cayman Islands- Frank Krasovec

Damselfish, for instance tend, to live in large groups. Most of the smaller species, like the lemon damsel, will feed and breed together as casually as the nassau grouper; leaving their young to the mercy of the tide. While the damselfish utilize colour to distinguish one another, most of that colour is invisible to the naked human eye. A spiral of different ultraviolet patterns is unseen except by specialist cameras and the fish's own eyes. Each pattern is the stamp of just one species and use of these advanced cameras is showing us how damselfish are able to distinguish their own kind from near identical relatives.

Larger damselfish species tend to make a higher investment in their young. The male sergeant major after mating, keeps watch over as many as two hundred thousand blue eggs. The eggs usually cover a rock and are so tightly packed together that they resemble an algal growth. Due to that they may well be overlooked by passing predators. Even so the male sergeant major attends to the eggs constantly, turning them over with his mouth and airing them with his fins. The male's prominent pattern is effective in identifying his aggressive nature when out in the open but as a last resort he can still withdraw and camouflage relatively easily with his surroundings.

The garibaldi is the largest of all damselfish and can afford to stand out. The lively scarlet fish can be heard roaring through the kelp proclaiming his territory. Rather than wander around the seaweed endlessly looking for a meal the garibaldi takes matters into its own fins. It is an avid gardener and plucks out any inedible algae it finds among the kelp to make room for more suitable foliage. But not all the algae it cultivates are for food. In the breeding season the male leaves great strands of red algae untouched. He also turns on any other creatures entering this protected zone, grasping starfish by their limbs and dropping them out of reach. The algae patch soon attracts females and as they draw closer the male begins bellowing at them, using his swim bladder to amplify the sound. He is as picky with his choice of mate as he is with algae and he begins to nip at her, chasing her fiercely around his garden patch and testing her strength. Once confident in his choice of mate he begins to bounce up and down repeatedly as if he was attempting the high jump. Once satisfied with each other the garibaldis settle down to the business of spawning in the red algae but it will be the male's sole responsibility to watch the young.

If the garibaldi male is dominant enough he may attract several females and eggs of different ages. In that case he may be left on guard duty for several weeks, supplementing his hunger with an occasional quick snack of some of the young. Other damselfish species leave their young where they will be safe from the appetite of watchful males.

Adult domino damselfish are dull grey coral dwellers, about as different visually from the garibaldi as possible. Their young however are dramatically patterned in black and white and require somewhere safe to hide. The shelter they choose is an unusual one, an anemone. Just like their close relatives, the jellyfish, anemones are covered with stinging polyps. A single accidental slide against one is enough to paralyse most

Garibaldi patrolling the kelp, Catalina Island, California- *Joe Belanger*
Sebae Clownfish cleaning eggs, Tulamben region of Indonesia- *Joe Belanger*

fish. The anemone does not sting itself; it recognizes the chemical make-up of its own polyps and thus avoids wasting its stings needlessly. As the domino damselfish approaches it cautiously slides against an outside polyp; gathering some of the invertebrate's protective chemical onto thick mucus covering its skin. It becomes more adventurous, moving like a dancer rapidly through the anemone until it has acquired enough of the protective chemicals that the invertebrate ignores it. The anemone will be its temporary home until it is large enough to look after itself.

One group of damselfish spend their entire lives sheltering among the anemones. Their safe refuge means they can afford to really stand out, even when compared to other damselfish. So vibrant are these anemone lovers they are all called clownfish, even though some of them are actually more closely related to other damselfish species than one another. Pattern and colouration vary mildly across the group; most of them are orange with broad white stripes. The skunk clownfish has one of the simplest patterns with a single slender stripe tracing its backbone. The pink skunk clownfish has elaborated this with an extra vertical stripe behind its head. The Sebae clownfish is distinguished less by its stripes and more from the dark ebony colour of its body, with the exemption of its orange nose. Most clownfish have developed the white vertical stripes on their body so extensively that they have lost the single dorsal stripe that may have originally characterised the group; it is lost amongst several other stripes and patterns.

One clown fish stands out among the rest. The maroon Clown is unmistakable not just in colour but in size. A female may be 15 centimetres long, the same length as a household ruler. Her stripes are not white but fluorescent yellow. Yet only the female maroon clownfish is this dramatic. The male is not even half her size. Maroon clownfish live in a rigid household of just two adults, with just the henpecked male and the dominant female. Any intruding clownfish will quickly be driven away from the anemone, generally by the female. Other clownfish, including the Sebae clownfish, live in larger groups, where individual roles are not as easily defined. This has created an unusual social dynamic.

The tomato clownfish may share its home anemones with five or even seven other clowns. Living in such large groups for an animal that survives equally well in pairs requires a strong social discipline. The clownfish society is maintained by fasting. The single female in the group maintains a typical dominance over the others using her bulk. If any of the males begin to approach her size she views them as a threat to her authority and vigorously drives them away. Once evicted a male is unlikely to find an unoccupied anemone and will soon perish. The lead male is equally insecure about being challenged and will drive away any males that get too close to his size. So, each male starves himself to remain on the anemone. The alpha male is no more than 80% of the size of his mate; he may even ignore a meal to achieve this. In turn the beta male will be less than 80% the size of his senior and so on through the entire group. Waiting outside the colony is a host of juveniles eager to join the hierarchy as soon as a vacancy becomes apparent.

Holding the community together is the offspring that the dominant pair shares. Sticking safely onto the base of the anemone is a cluster of eggs which the alpha male constantly cleans in his mouth and ventilates with his fins. It is his skills at parenthood which will determine whether he too is chased away from his home. And there is a host of predators that would relish a snack of clownfish eggs. Tomato clownfish are forced to guard both eggs and anemone from creatures as large as hawksbill turtles and sea snakes. Even the mother will join in this defence. But being such a vigilant protector has its risks and occasionally a female will be killed. The death reverberates through the small makeshift family. Neighbouring females have little interest in taking on defence of another group and the previously dominated lead male is left on his own to defend both his young and territory. Nor does there appear to be any reprieve to be found in the juveniles competing for a spot on the anemone. They are all male. And the alpha male has no more courtesy in sharing his property than his deceased mate did.

Without the watchful eye of his mate the lead male begins an uncontrollable glut, gulping down whatever food he can find. He grows drastically, soon reaching the size of his deceased partner. But the clownfish appears anything but content. His personality changes, he starts to turn on his companions in the anemone, not just occasionally but continually. The eggs he once guarded so proudly may be discarded or even consumed. The decline is not only behavioural but physical. As the male's erratic behaviour continues his gonads begin to shrivel until he is no longer able to be recognised as male. If a female does appear this debilitated fish will certainly not be able to fertilize her.

Despite his active neglect a few of the clownfish's eggs may have survived long enough to hatch. There is little consolation to be had. Again all of the infants darting among the anemones are male. And it is a situation that can be found right across the reef. At first sight the entire population appears at risk of extinction without the necessary number of females to maintain population. But there is more, quite literally, than meets the eye. At last the male stabilizes and the extent of his mutations is revealed. The emasculation of the formerly hen-pecked male is complete. He has become a female. He initiates courtship with one of his former competitors. Together they can now ensure the continuation of tomato clownfish on the reef. Other clownfish, including the mighty maroon species, also have the same reproductive strategy. The clownfish is one of the few backboned animals where mature males can turn into females.

Bringing forth the next generation is as acute a challenge in freshwater as it is in marine environments. In the British summer stickleback males begin to flash bright red from their fins as they prepare to breed. They start to build nests, stripping reeds with their mouth and gathering it into shape. Such work isn't easy when you can only use your mouth so the males defend their nests proudly. They are also aware of the strong territorial nature of their rivals; when travelling in an opposing territory they quickly lower their fins and head for cover. Guarding the nest comes at a risk. If any predators approach the stickleback has only two options; to stay and potentially be eaten, or escape and leave its eggs unguarded. Such a dilemma is also faced by distant relatives of the stickleback; the seamoths.

These delicate marine creatures bear more than a passing resemblance to an elaborate doily. Their pattern also provides superb camouflage, allowing seamoths to slowly investigate the seafloor for shrimps and other crustaceans. There are five species; all found in the southern hemisphere. Some live in reefs, others in kelp but all of them are almost indistinguishable from their surroundings. To aid in the deception their body shape is distorted, removing any resemblance to a fish. A male seamoth has bright colours on his underside which he reveals to a female by quickly flicking the fins out of the way. Having gained her attention the two of them float together for a time, gradually rising above the seafloor together. In one dramatic moment the pair spawns, discarding their eggs to the whims of the ocean.

The seamoths are most interesting due to their genetic positioning between two groups which both invest heavily in their young, the sticklebacks and the pipefish. The pipefish order contains a number of varied families besides the main pipefish family.

Seamoth, Indonesia- *Joe Belanger*
Ornate Ghost Pipefish, Bali- *GreensandBlues*

Trumpetfish, for instance, sweep like a siphon through shoals of small fish, sucking several of them up at a time. Occasionally they can deploy their stealth; trailing behind a parrotfish and surprising any prey the larger fish disturbs. They can even be brazen enough to snatch prey out of a grouper's mouth.

The armour plated shrimpfish has a similar shape but targets smaller prey, chiefly shrimp. Unusually most of the time their body is vertical, with the head pointing towards the sand, they only change position briefly when swimming between feeding grounds. The fish shoal closely together, elegantly swaying with the current so that they resemble the spiny plants they frequent. They may even shelter in this way among sea urchins. When it comes to breeding the shrimpfish come together just as amiably to lay eggs which, like the seamoth, they then abandon. Many of their relatives however are significantly more attentive parents.

Ghostpipefish have bodies which are even more rigid than that of shrimpfish. Their star shaped bony plates provide efficient defence but also reduces speed and control significantly. Depending on the species, the combination of colour and awkward positioning of fins results in a resemblance to a stray piece of seaweed or even an inedible starfish. Arguably the most dramatic species is the ornate ghostpipefish with its surreal combination of yellow and indigo yet it disappears once it freezes against the coral. Its awkward body shape restricts it to relying on the inefficient flapping of its pectoral and pelvic fins to get around. Perhaps because of their poor swimming style ghostpipefish invest heavily in their young. After a pair spawn adhesive eggs are stuck to the female's pelvic fins until they hatch. The female can now only swim using her pectoral fins; her cryptic camouflage becomes even more crucial for her survival. Once hatched though her young will have to take care of themselves.

True pipefish also invest heavily in both camouflage and their young. While their body is still encased with bony plates these are positioned in such a way that the fish is still able to flex significantly. It relies almost entirely on its swaying body to swim and consequently has lost all fins but those around its chest and even they are minute. Pipefish resemble a drinking straw and have the same ability to bend the neck back to allow food into its mouth, their toothless narrow mouth restricts them to a diet of microscopic planktonic. They also have no stomach so any processing of food is completely inefficient. As a consequence they have to feed constantly, gulping down several thousand shrimp larvae a day. Feeding itself takes energy so pipefish have developed a way to stay near a food source without swimming. The tail has a flexible tip, allowing it to grasp. Several tree dwelling mammals and lizards also have a prehensile tail but pipefish and their close relatives are the only fish to have one, although it is not present in all species. The ability to stay firmly attached to one spot has allowed some relatives of pipefish to develop their camouflage to extreme levels.

The seadragons are found off the coasts of Southern and Eastern Australia. They are unusual in keeping their anal and pelvic fins but they are not restricted to routine locomotion. Instead, rippling from their trunk, fins and tail are peculiar flaps of waving skin that blends in with the kelp in which they shelter. The leafy seadragon resembles seaweed so closely that small fish might actually shelter in what appears to be just harmless foliage. The only other species of seadragon in the world, the weedy seadragon, by contract is a fascinating mismatch of colours that seem to mimic that of coral, even though this species actually lives in temperate waters including that around Sydney.

◄Weedy Seadragon male with eggs, Botany Bay, Australia- ***Harry Rosenthal***

Both species of seadragon spawn in pairs but the compatibility of this union has a higher level of importance than it does for most fish. They perform a courtship dance for several days until they feel confident to take their relationship further. When both feel comfortable the female draws near to her partner and slowly begins releasing eggs. A few eggs will be washed away but those that survive must stick to either the male's tail or his abdomen. Once the eggs are attached the male fertilises them and his body takes on a physical change. Skin bubbles up around the eggs, forming individual caps that hold them in place. Now no matter how dense the kelp is that the male passes through his eggs will remain firmly contained and will stay there for a month until hatching. The female during this time is restoring her energy reserves enough to produces the next batch of eggs and as soon as one brood of seadragons hatch she passes the new generation over to the male. Due to their parents' strong bond the seadragon young have a higher rate of juvenile survival than they would otherwise.

A more typical pipefish, the banded messmate, also keep the eggs stuck to the male's underside, but take the parental care one step further; they have a pouch. Once the female deposits her eggs within it the sperm duct empties into the pouch and fertilizes the offspring. As the eggs grow hormones begin acting upon the male allowing him to experience the full state of pregnancy. He begins pumping extra nutrients into the pouch to nourish his young. With this additional food supply his young mature quicker and leave their father after a few weeks.

The majority of pipefish species are, like the seadragons, monogamous although a few are ruthless philanderers. The gulf pipefish male is typically loyal to his mate. She is after all larger than him and has fought quite competitively to win him from the other females. But occasionally a male may end up with a female that is smaller than him and likely to produce weak young. He has little choice in his partner but his body plays a devious trick if he is uncomfortable with the union. When the female deposits her eggs into his pouch they will not be fertilized. The pouch ceases acting as a cosy nursery; indeed it actually begins to absorb nutrients out of the eggs. It has become, so to speak, a second mouth. Only when the male gulf pipefish has a mate large enough to produce a sufficient quantity of eggs will he finally undergo the process of pregnancy and delivery; until then his body will treat each mating as an opportunity for an easy meal.

Pipefish have close relatives that are even better known for the behaviour of fathers to care for their young in pouches. One archaic group of pipefish restricted to southern Australia and Tasmania, the pygmy pipehorses, give an indication of how these intriguing species evolved. Pipehorses have a body similar to other pipefish, excepting their equine shaped head. In horses this solid shape enables the mammal's skull to withstand the force required to constantly grind away at brittle blades of grass, in the pipehorse the external force they need to withstand is being driven at speed through the water. A pipehorse will move ever so imperceptibly towards a potential meal, remaining just far enough way that its target, such as a small shrimp, isn't triggered to move further away. What the shrimp isn't expecting though is for the head of the pipehorse to slam through the water, accelerating towards it with such speed that the shrimp goes from being free to being consumed in less than the fraction of a second. Having such a bulky head though considerably slows down the swimming ability of the pipehorse, restricting them to algae based kelp forests, where their adaptation to cooler waters has allowed them to avoid more ruthless competition, and predators.

A massive environmental change between twenty and thirty million years ago provided the opportunity for close relatives of the pipehorse to escape the cooler latitudes and eventually reach the tropical riches of the reefs. Continental movements and sea level change lead to the creation of massive shallow coastal expanses, in latitudes too cold for coral, but warm enough to support a comparatively recent terrestrial invader, sea grass. The sea grass provided ideal form of shelter but they also provide multiple support frames for those able to grip onto them. Seahorses are experts at gripping, indeed conservationists at some facilities will allow visitors to place their hands in the water and it doesn't take long for one of the enterprising swimmers to take advantage of the newly available perches and wrap their tail around a finger. They mightn't have to keep swimming while holding onto the seagrass but they are far from resting. If you watch them closely they can be seen to be feeding continuously on miniature invertebrates, using the anchored tail to help the body nice and stationary, allowing the head to take full advantage of the stored kinetic energy and drive itself towards its prey at ease. They are surprisingly effective predators; efficient and deadly to anything small enough to be caught within their jaws. Their adaptability has allowed seahorses to leave their original seagrass habitat and to live among the coral, though even then they are still happy to cling their tails around any seagrass or kelp they can find.

Their highly specialised adaptation means that seahorses lack the ability of pipefish to flit easily from one environment to another. Indeed, their swimming is even poorer than the seadragons and pipehorses; rather than move their whole body from side to side most of their motion is generated from an oversized fin on their back, they tend to prefer clinging with their tails onto a piece of seaweed or coral rather than actively swimming. The cause of their stiffness is bony armour that encloses their entire body to the extent that they lack scales and, even more peculiarly, ribs; the armour is sufficient to protect the chest cavity, and the fact that they breathe exclusively through gills rather than lungs means that the chest does not have to be able to expand in order for them to breathe.

Seahorses feed relentlessly on small shrimps and their inability to chase after them means that each section of coral can only support a few individuals. Consequently, rivals of the same sex engage in a ritual battle whenever they encounter each other. Pygmy seahorses use their heavily enforced skulls to ram each other across the flanks. One of the pygmy seahorses, Bargibant's seahorse, is so small that up to 28 individuals can be found on a single seafan, and their territories overlap, meaning that it does not take long for a seahorse to become embroiled in another squabble with its neighbour. The vast territories of the narrow-bellied seahorse, by contrast, mean that there is a much lower chance of coming across another individual of the same species. Not only does that reduce the level of fighting it also significantly influences their reproductive choices; most narrow-bellied seahorses have just the one mate for each breeding season, and their isolation from each other, means that some individuals may continue pairing with each other for multiple seasons.

For many years monogamy was believed to be the norm for all seahorse species. Detailed observations in aquariums of several hundred individuals of different species have shown more of a nuance in the social and reproductive behaviour than could previously be glimpsed over comparatively short periods of time at sea. Some seahorses regularly mate with seahorses that are not their primary partner, and same sex pairing occurs occasionally with female seahorses and regularly with males.

Seahorse- ***AddictiveStock***
Baby Lined Seahorse, captive bred from wild caught at Cape Charles, Vancouver- ***Chris Crippen***

On very rare occasions a trio of seahorses has even been observed together. The environmental and behavioural pressures underpinning this variety of behaviour are less well understand; does engaging in non-reproductive mating reduce the level of aggression and resources used by seahorses which would otherwise be competing with each other, or, does reducing the number of potentially successful breeding attempts mean that there is more potential energy available when reproductively compatable mating does occur?

The behaviour of a breeding pair of seahorses is a little better understood. One of the most studied is the dwarf seahorse, although curiously enough this species is slightly larger than pygmy seahorses. Being small and with their key defences being mainly camouflage and their lightly armoured bodies there is no guarantee that when they wake up the next day that their partner will still be there.

In some cases the relationship between pairs is so strong that if one of them disappears, the remaining seahorse may never seek another partner. Perhaps as result of this close bond each morning begins with a greeting dance that lasts around ten minutes, the pair lifting up and effortlessly mimicking the other's fin strokes, sometimes with their tails wrapped closely together as they do so. Synchronised movement is an essential component of their long term relationship. After giving birth, for instance, the pair reinforces their bonds with a full courtship ritual that may last nine hours, an incredible feat for an animal that needs to eat so frequently.

A courtship ritual of such extraordinary length is not entered into lighty. Some days a pair, particularly a new one, may only enter the first of the four phases of courtship and continue no further. Initially it can be difficult to tell that the courtship has even started. Two dwarf seahorses sidle up to each other, ever so cautiously sliding their tail next to the other until it is so close that it is almost, but not quite, touching. A male turns his head to the female and does a slight shimmering of his head, almost imperceptible if you don't know what to look for she. She copies. If you look closer it becomes clear that the skin actually changes colour as they shimmer, the lack of scales allowing them to easily flaunt their glowing skin at their partner. This can continue multiple times, occasionally accompanied by a change of colour. After several days of engaging in this, if the couple feels compatable, they then move to the second phase of their courtship. The female takes the lead now, leaning straight over towards the male, and holding her position. The male actually leans *away* from her and gives another shimmer in responce. That second phase can last nearly an hour. Holding those extended positions pushes their stamina, and their bodies, to its physical limit, and they have a rest. The only indication to a passing human observer during the seahorses much needed interval that they are still mid-courtship is the male seahorse's occasional habit during this of pumping his body impatiently up and down, like some an overly inpatient love-sick puppy. When they have rested enough to commence the third phase the female resumes leaning over towards the male, glowing brightly as she does so, and the male simultaneously glowing and leaning back in return. That phase can be over and done with in less than ten minutes, although it can keep repeating another five times. At last they leave the security of the perch for their fourth stage; back fin beating wildly, tails finally entwined together, preventing their potential separation from each other mid-swim. Another species, the White's seahorse, takes this opportunity to demonstrate to the female his availability and peak health; inflating his pouch to its maximum volume than opening it wide so that she can see he is not already caring for the brood of another female. Satisfied with her inspection she is ready to conclude the courtship. Only now will the pair feel comfortable enough for the female to release her eggs into the male's pouch and for him to finally fertilise them.

They take such care and engage in such elaborate bonding rituals because for both sexes bringing forth the next generation is a major investment. If the eggs, for instance, are not released before the male's body is in sufficient condition than he is unlikely to bring many of them to full term. Perhaps their extended dancing sessions provides an opportunity to allow both seahorses to assess the other's condition before they commit to breeding, perhaps not. It may also allow them to precisely identify their chosen partner and somewhat avoid the likelihood of inadvertently breeding with one of the neighbours from one of the overlapping territories, instead. What is clearer is that when the female does release her eggs into the mle's pouch, they can make up to thirty percent of her body weight; her priority now is urgent refueling. For the male it is his time to put his body on the line for the next generation.

Just as in the pipefish nutrients are redirected from the rest of the father's body towards the pouch. As his partner begins to slowly recover her lost weight his own body is increasingly subjected to the energy-sapping processes of pregnancy, although studies measuring oxygen consumption before and during this reveal that his role is considerably less energy draining than producing eggs is for the female. It takes about a fortnight after conception for the baby seahorses to be mature enough to leave the safety of their father's body. As the pouch slits open the departing offspring are perfect miniatures of their parents. From the moment of birth their eyes are open and their tail muscles strong enough to wrap around a branch of coral. And they need to be. As soon as they've left their father any parental link is also broken; not only will the adults ignore their infants feeding requirements they may even make a quick snack of them to recuperate their energy. Some species of seahorse allow their young to take shelter in the pouch but these are the exception not the rule. There are, however, completely unrelated spiny ray finned fish that actively care for their young.

Like seahorses the leaf fish spend most of their life safely camouflaged; only in this case hiding amongst the tangle of weeds in streams and stamps in Peru. Their name is fittingly apt; the only feature likely to make them stand out to human eyes amongst a heap of other submerged rotting debris is that their leaf shaped body is permanently held at a vertical position. These fish are so well disguised that their prey may even enter their mouths, fooled into thinking the motionless predators are harmless pieces of foliage. But leaf fish eggs do not have the same deceptive advantage; nor are they capable of eating anything that would like to take a bite of them. To survive they require the active protection of their parents. While the eggs develop the fish keep a watchful eye; depending on the size of potential threats they can either chase or consume them.

Juvenile shrimp gobies also rely heavily on their parents. Shrimp gobies are vulnerable not just because of their miniscule size but also due to living in barren stretches of seafloor beyond the coral reef. Here they would be easy prey were it not for their association with a half-blind crustacean that shares the same waters. Pistol shrimps dig burrows beneath the sand for shelter but to find fresh food they still have to venture outside. Their short sightedness severely impairs them but by sharing their residence with the gobies they acquire a virtual guide dog. The small fish stay within reach of the shrimp's antenna at all times, the moment it departs the shrimp automatically returns to the shared hole. The burrows provide effective protection not just for shrimp gobies but their young. The sleeper goby, in particular, exploits the full potential of the burrow.

Shrimp goby- *Xcldhx038a82fd*
Southern Cardinalfish mouth brooding, Mornington Peninsula, Victoria, Australia- *Corey Doughty*

This beautiful orange shrimp goby stays together in close knit pairs. Most of the time the pair barely moves apart but after the female lays eggs at the end of the burrow the relationship changes dynamically. The male begins making his own excavations but they are for an entirely different purpose than those of his shrimp flat mates. The female moves into the new extensions while the male starts dragging soil to its entrance; locking her away from the rest of the burrow. The attentive male will even drive off any shrimp that attempt to clear the barricade. Eventually the female will be cut off entirely, living off her fat reserves as she stays with her eggs. She may be hungry but she is also safe from any passing predators. It also protects the eggs from being eaten inadvertently by the shrimp.

The female sleeper goby is hidden but she is not forgotten, at least not by the male. He continues checking the wall separating them until he can hear the movement of little fins swishing around. Only now will the male sleeper goby allow the shrimp to dig through the wall, releasing the female, where she returns, rather rapidly to resuming her feeding outside the burrow.

Another protective parent, the jawfish, can be found in tropical shores across the globe. It too hides in a burrow but unlike shrimp gobies it has to make one from scratch; instead using its vast jaws to scoop out its home, often unearthing suitable prey during the process. This task occupies most of the day but the burrow is only large enough for one adult. The only available shelter for jawfish eggs is within the mouth of their father. This does cause considerable problems though when it comes to feeding. The jawfish could let the larvae out of his mouth whenever he saw a meal but the crustaceans he feeds on are similar in size to his young. Instead the father astutely keeps his mouth closed as his babies develop inside it. It could hardly be the most comfortable experience; the eggs filling his mouth are so large that his cheeks visibly bulge and he can barely keeps his jaws together. Eventually he opens his mouth to let his half formed infants float out to forge their own independant lives under the soil.

Other small reef living fish also shelter their young in their mouth, cardinal fish for instance. Like the jawfish their mouths are crammed full of eggs, but their fragile jaws mean they are unable to dig protective burrows. Instead some of them find shelter by darting within the spines of sea urchins where few fish will dare to threaten them. One nocturnal species, the southern cardinalfish, has found a way to multitask between parenting and looking after itself. Rather than restricting itself to living alongside a sea urchin it hides among cracks and crannies on the reef's edge, allowing a mouth brooding father access to potential sanctuary whenever potential threats approach. He is also shadowed by the mother of the eggs, who is quick to defend him and drive off any competition in his vulnerable state. Standing guard also allows her to keep an eye on him so that he isn't tempted to swallow a few eggs though, just to be on the safe side, the first eggs she produces tend to be decoys (just oversized otherwise normal cells) and these are the ones furthest down his throat. Even after the young hatch the male's responsibilities do not cease entirely, they will keep returning to his mouth for shelter while they still fit. But the father's appetite is slowly becoming a more urgent priority, so, rather than continue to ignore any particularly appetising prey, he finds a suitably sized gap in the rocks and spits all the juveniles into it. Only after a decent feed will he be ready to scoff up a mouthful of his young without swallowing any of them- most of the time. It is not without reason that the group of cardinalfish that this species belongs to has the rather unflattering name of gobbleguts.

Perhaps the most devoted of all fish parents are found not in the sea but in fresh water, the cichlids. Physically cichlids are notable for the additional set of teeth in their pharynx; which allows them to process different food to their regular teeth in the jaws. In India and Sri Lanka are two species of cichlid, the orange and the green chromide, which are among the most primitive of the family. The only visual difference between the sexes is that the male is slightly larger. For such indistinct fish they take special care when it comes to choosing a partner. After completing such proceedings they spawn in any gaps between rocks or submerged branches and belligerently guard their eggs until they hatch. This relatively simple behaviour has been adapted by other cichlids to produce some of the most complex parenting provided by any fish.

The diversity of cichlids is displayed at its greatest in the vast rift valley of East Africa. Here a network of three major lakes and their interconnecting rivulets provides a home to several thousand species of cichlid. In Lake Tanganyika a male shell brooding cichlid is preparing to mate and, as his name suggests, finding a suitable shell is crucial to this. Such a demand might seem simple to fulfil for there are many snails munching their way through clumps of algae. But the majority of shells are too small to hold many eggs and have to be ignored. Other shells nearly overshadow the male's girth, yet he stoically tries to claim one of them. Shells of the right size are valuable real estate and there are many fights as the cichlids try to prove the rightful claimant of the property. Once a male secures a shell he is forced to drag it to his territory, often fighting with other males as he inadvertently passes through their territories in the process. When the diminutive female appears the purpose of these oversized shells becomes apparent. The shell has to house her as well as her eggs and a smaller shell just wouldn't fit her. Her partner's strength will continue to be tested over the coming weeks as he stands guard outside, chasing away anything that could threaten his family. These shell brooding males are so efficient at collecting large shells that in the largest harems fourteen different females may be gathered together under the watchful eye of just one male.

Blue Tilapia cichlid building a nest, Florida- *Alek Tritt*

In the other major African lakes cichlids have developed equally elaborate pre-mating rituals. Lake Victoria previously had the most varied and unusual populations of cichlids but the introduction of foreign predators has removed much of their number. The third major lake in the rift valley, Lake Malawi, is still filled with hundreds of species, each of them slightly different in appearance. There is another major distinguishing feature that varies between species, the shelter which the male provides his mate. There are scarcely enough snail shells to hide vulnerable mothers nor is there a surplus of tangling tree roots dangling into the lake. But there is a surplus of soft soil and most of the cichlids here use that to build their own constructions. They begin taking mouthfuls of soil and spitting it out to create a circular mound. In some species this surrounding mound becomes the main feature and it may outline the male's territory; resulting in a courtship arena with a diameter of a metre of more. Others put their effort into prioritising the excavation instead, creating a smooth saucer like effect that spreads for over three metres. In any lake, even one as large as Malawi, there is only limited room for these courtship arenas so they are often just inches from each other. When some species come together in leks there may be several thousand of these constructions massed together.

The curious female investigates and enters the territory of whichever male impresses her the most. He trails her, flicking his fins impatiently until the moment she releases her eggs. Then the technique varies. In some cichlid species females will allow the male to spawn and guard the eggs, and later the developing young, inside the mound. Others find it a little harder to part with their young. They protectively swallow the eggs, often before the male even has a chance to fertilise them. But the male is prepared for this. Decorating his anal fins is a series of golden spheres that resemble yet more eggs. The female lunges for the spheres, gobbling the sperm that her partner releases into her egg crammed mouth. Her lips will remain closed for the next couple of weeks until her young hatch. Some species will release their young into the dome that their partner has so diligently constructed but for others even that isn't enough. Cichlid larvae are born with an excess of yolk in their stomachs but this only lasts them so long. Eventually they have to leave their mother's mouth to feed; exposing them to the perils of the outside world. But even now she does not abandon them. Instead, often both parents will accompany the young, ferrying them between feeding sites. Often the adults' size is enough to deter most larvae predators. But if a threat remains the parents have one last defence strategy; allowing their young to return to the shelter of their jaws.

Cichlids also live in South America. The oscar is one of the largest cichlids, reaching eighty centimetres long. This giant is not particularly careful of its eggs; it simply deposits them on a leaf and leaves. The young have bright orange stripes and dart through the waters but as they age they assume their parents sombre grey appearance and personality. The Oscar has however a close relative, graceful and distinctive enough to have its own following among aquarists, the discus. This Amazonian fish is shaped like a saucer and slips through the weeds picking off tetras with such grace that they barely notice it, until it's too late. The discus moves at a slow pace and is unable to outswim either predators or prey; instead it relies on its camouflage to outwit both.

Adult discus may appear fiercely independent but they have to rely on each other to successfully bring up their young. During adolescence these delicate cichlids begin their courtship process; each discus displaying its fins and gently circling each other. No mating occurs yet but it does establish the compatibility of the pair. Once a pair is established they begin stripping a large rock of its algae and the female lays her eggs over the cleared area. New parents seem uncertain and the unattended eggs become overgrown with algae and wasted. The naïve parent responds by eating what is left unspoilt. A few weeks later the pair may attempt to spawn again. This time the adults may have learnt to aerate the eggs with their fins and tend them with their mouth. After devoting several days to this purpose they become overexcited when cleaning the eggs and consume the lot. Such misfortunes will continue several times for the young cichlids. It may take ten attempts until any young hatch for the pair. Even then their work is not finished.

Juvenile discus can't feed as efficiently as their parents and risk being eaten themselves if they were to attempt to hunt as adults do. Fortunately for them the parents provide food. Rather than shepherd their young between feeding grounds the adults' bodies begins to produce a specialised form of mucus from the skin. The mucus is high in both nutrients and antibodies and assists the young's growth as well as boosting the immune system. The discus is effectively feeding its offspring on milk. Both parents produce this substance and the growth and survival rate of infants that grow up with their parents is highly improved from those, such as in some aquariums, that do not. Discus are born with the standard stocky fish body shape but as they grow their body stretches vertically, making them harder to swallow. Other ray finned fish show even more extreme changes.

Flounder- ***Photo by Xcldhx038a82fd***

Flat fish (creatures like plaice, turbot, halibut and flounder) start life looking like fairly typical larvae. When they hatch they live with the rest of plankton at the surface but after six weeks their body begins to develop to fit their adult life on the seafloor. Many fish that live here, such as rays, have their bodies flattened diagonally so they can simply lie flat on top of the sand. But flatfish are flattened vertically just like discus; which is quite an awkward shape considering its lifestyle; they will spend their adult life permanently lying on one side. In larvae of the adalah, one of the most primitive of the group, the strangest change is that both eyes gradually move closer together until one pops up on the other side of the head. With this transition complete the adalah plops onto the substrate, able to observe potential food and threats with both eyes. The mouth also changes position but only subtly, restricting its ability to catch and swallow food. Peculiarly the speed of change is sex-related, male adalah develop into adults much quicker; a feature that so far seems to be shared among all flatfish, although future discoveries could possibly change that.

Adalah are unusual in that either eye can make the transition to the other side of the body. In most types of flatfish only one eye is able to move and that is genetically determined. Indeed, two of the families of flounder (and there are multiple) are separated simply by whether it is the left or right eye that moves. In these cases the migration of the eye is so significant that both eyes line up evenly in the middle of the head, so that it comes resembling any other fish, albeit a significantly squashed one, until you realise the other side of the face has no eyes at all. The mouth too has moved so that it can better handle a passing meal.

Flatfish are some of the most surprisingly diverse groups of spiny ray finned fish; by 2025 there were twenty different families- one of them only added that year. There is indeed some disagreement whether it is truly a natural group or perhaps contains similar looking descendants from completely unrelated families. Regardless, that distinctive lopsided shape, regardless of how awkward it appears to us, is proving itself a highly successful one. One of the flatfish, the smallhead sole, undergoes one of the most visually dramatic transformations of the entire group. From a simple transparent larva its skin becomes brightly patterned unlike any of its well camouflaged relatives. Its main body is a mustard brown with stripes running across it that a tiger would envy, with a vivid blue border; although there are a number of different patterns including a black body with an electric orange fringe. It stands out deliberately. The small head sole is one of the marine world's great impersonators. Its striking appearance and peculiar scurrying wriggle across the bottom allow it to resemble a poisonous sea slug found in the same area. But such a disguise is only effective when the sole's predators can see it. At night the small head sole has no option but to hide beneath the rocks while the coral reef is handed over to the true masters of the darkness.

Chapter 7. Denizens of the Deep

Onefin Flashlightfish, Fiji- *Mark Rosentein*

As the sun sets on the Great Barrier Reef the multitude of fish feasting in its tropical waters seek shelter. As visibility drops any unprepared fish in the open risks attack from sharks. Garden eels withdraw into holes, sealing their burrows with a pad of thickened saliva and sand. Clownfish sleep secure inside their anemones. Parrotfish create the most extraordinary of sanctuaries by blowing a bubble of slime, sealing both them and their scent securely within. To do this they slowly pump air exhaled from their gills and mouth back beneath the slime that naturally occurs on their skin, forcing it to expand. Completing the process may take a good hour. Once hidden in the bubble the parrotfish remains motionless, drifting in and out of sleep until the returning sun can give them confidence once more.

The coral itself provides a temporary protection for many smaller species but larger fish have to look elsewhere. The few caves sunken into the headland provide some protection. Sunken wrecks also provide an ideal sanctuary but to reach them means swimming across open water. White tipped reef sharks patrol the darkening waters and are quick to snatch any fresh prey. But ship wrecks and caves can harbour their own equally efficient hunters.

While their competitors have been feeding during the day squirrelfish have been sleeping inconspicuously in the darkness. As that darkness spreads they leave the wreck, relying on the brief trace of moonlight flickering through the waves to navigate. Their eyes project like orbs from the sides of their heads and are able to take advantage of even this mild light. These eyes are their only conspicuous feature under natural light, everything else blends in with the darkness. But a torch reveals that squirrelfish are actually vivid red. Surprisingly this dramatic colour actually acts as camouflage due to the unusual effects of water on light. Unfiltered light directly from the sun appears white. But the density of water actually splits the light into different wave lengths. Red light is quickly absorbed and even a few metres below the surface eyes strain to see the colour. Below fifty metres anything red is basically invisible. At night such a strategy is equally effective and the squirrelfish travel right to the surface of the water to harvest plankton. Like bats these fish make a dramatic exodus from the caves each night as weary diurnal species flock towards the rapidly vacating shelter.

Even with such bulging eyes squirrelfish struggle to see over long distances and rely on something else to keep their shoals together. Like parrotfish and cichlids they have additional rows of teeth in their pharynx; which can be used for more than just feeding. By grinding these together the squirrelfish are able to smoothly communicate and identify each other's whereabouts. A layer of rigid scales across their scarlet bodies provides some defence, although not much. Other nocturnal fish have both more efficient protection and vision.

Off the coasts of Japan a curious lightshow of aqua green beams begins to play each night. The light is being radiated by bacteria sheltering together in their billions, although curiously under artificial light they appear no longer green but orange. Equally interesting is the place they have chosen to shelter; below the eyes of a pinecone fish. These fish have pouches at the tip of their lower jaws that both house the bacteria and provide a source of nutrients to maintain them. The luminescence draws inquisitive shrimp and other plankton directly to the mouth. It also attracts predators large enough to threaten the pinecone fish. Its rigid scales are virtually impenetrable and the spines projecting from its pelvic and dorsal fins deter most hunters. The combination of both armour and the rigid spines restricts both of these sets of fins and overall flexibility, forcing the pinecone fish to rely almost entirely on its spinefree pectoral fins for swimming. Having such efficient armour plating is enough to deter most predators and so the pinecone fish maintains its niche quite comfortably, not only in Japan but also in Australia. Here it comes in contact with a relative with vastly improved vision adaptations.

Australian Pineapplefish, Port Hacking, Sydney, Australia-*Harry Rosenthal*

Pineapple fish are also heavily armoured but it is their eye pouches that sets them apart from their cousin. These are positioned directly below the lower jaw, rather than being restricted to the end, and because of their position they can afford to be larger and consequently provide more light. Not only can prey see it but the pouches are bright enough that the fish can actively track down and pursue a meal. But having a permanent light source has its disadvantages, it is impossible to sneak up on prey or to hide inconspicuously from predators like moray eels which can pierce its skin. Without a way to conceal itself the pineapple fish has no option but to rely on its armour alone, and that slows it down considerably.

Flashlight fish also have bacteria filled pouches, in this case located below their eyes. The bacteria are even more concentrated, ten billion squashed together in each cubic centimetre. While these fish are unable to control their light source any more than pineapple fish can they are able to disguise it. A large flap of skin lies beneath each pouch which can be brought up like a screen over the bacteria. Shoals of flashlight fish glide over a reef, rhythmically flashing up areas of it, before vanishing. Because of its ability to conceal itself it is able to rely on speed rather than armour for protection. When they sense an incoming shark they stop flashing and quickly dart off in a different direction. They can even use their lights to warn the rest of the school of an approaching threat. Alternatively, they may inadvertently flash their spotlights directly on their competitors only to see them be eaten up.

Further out to sea the setting sun is the cue for the largest migration on earth, one that out-scales even that of sardines or wildebeest. Each night several billion tonnes of fish and other marine life swim up to feast on plankton. In the dark they are less vulnerable to many predators and they gorge till content. They do not feed blindly. Lantern fish, one of the most numerous fish, don't rely on bacteria but instead they produce their own light by a chemical reaction called lucerifin, occurring between their skin and the surrounding water. Similar to the reaction when a chemical is added to a Bunsen burner a bright light is produced, in this case white. Each species of lantern fish has a slightly different arrangement of photophores, light producing cells, which allows them to tell each other apart at breeding time. Indeed, a single existing species of lantern fish, which truly is the exception that proves the rule, completely lacks photophores allowing it to easily recognise each other and spawn successfully. These different patterns also allow them to recognise the distinctive markings of the specialist hunters tracking light fish up from the deep as many of these also have photophores.

Armed with specialist eyes and oversized teeth most of these specialist predators are more than capable of devouring a fish their own size or larger. One of these, the bristlemouth, is the most numerous fish, and indeed backboned animal, on the planet. As the sunlight returns it becomes vulnerable itself and has to descend, along with the shoals of lanternfish and other nocturnal feeders. The average maximum dive for a person is less than fifty metres but the lanternfish easily outdistances that. Below two hundred metres it is too dark for coral to grow and the rock wall is bare but still the fish continue. The deepest human dive ever recorded, without highly specialised equipment, was down to a hundred and thirty three metres. After this a human simply has to return to the surface to breathe and will likely suffer the debilitating effects of decompression sickness as a result of their efforts. To follow the lantern fish all the way back to their daytime shelter requires a submarine or a ROV (remote operated vehicle.) Below a kilometre most of the lanternfish have scattered, as the reduced plankton doesn't support large groups. Although many different groups of animals, including squid and whales, undergo these nightly migrations no others travel as far as the lanternfish each day. It takes them three hours to reach their final destination. For some species this may be as far as 1,700 metres below the surface. They have swiftly outstripped many of the hunters chasing them. But in the depths there are local predators that could also make quick work of the lanternfish.

Fangtooth are some of the most common. The juveniles can be found closer to the surface and are notable for the spines sticking up above the head but these defences are slowly absorbed as the fish grows. As they grow the head also becomes more prominent while its skin increasingly darkens. They also begin to travel deeper, stopping after they get about a kilometre or so below the waves. By the time they have reached this depth they have morphed into a fearsome predator worthy of their name, a dystopian looking creature that could be mistaken for the misshapen love-child of a bulldog and a tin can.

When it comes to the adult fangtooth you literally do see their fangs first; indeed their teeth swell out of their jaws. These formidable looking hunters have some of the largest teeth per body size ratio of any animal. In order to close its mouth properly its skull has holes penetrating into it that its bottom teeth can slip into. With a head that makes up nearly half of their body fangtooth are fully equipped to tackle any small prey that comes their way. Speed also makes them one of the top predators of their depth. So their size then can come as something of a surprise, roughly 15 cm, about half the length of a standard ruler. Any particularly large lanternfish will have to be ignored. The deep sea simply doesn't provide enough regular nourishment for large predators so consequently few of them exist.

Spookfish have a totally different survival strategy. Their mouths are tiny, just a tube to suck up little copepods. Spookfish are distantly related to salmon but different living conditions have completely changed its shape. The sheer pressure of water on its body continually crushes against its body. In the jelly faced spookfish this has resulted in an awkward round shape which is structurally strong. It can't swim very fast but nor can it be easily swallowed. Great eyes balloon from the jelly faced spookfish's face. These are highly sensitive to any changes in light and can quickly spot the silhouette of potential prey floating above them. They can even distinguish between photophores and natural light. But these are not the most unusual eyes in these oceans, even among spookfish.

Fangtooth- ***Sandra Raredon, Smithsonian Institution***
Jelly faced Spookfish- ***Brauer, A., 1906***

The common spookfish is about as different in shape from its cousin as possible. Its body is peculiarly long, almost as if it has been stretched out on a medieval torture rack. The pectoral fins extend like delicate filaments that are about half the length of the entire fish. But once again it is the eyes which really distinguish it. These literally stick out like a pair of binoculars and provide a similar quality of vision. They share one serious flaw with manmade binoculars; by focusing straight ahead they can miss objects that are just to the side, like moving predators. So discreetly placed to the side of each cylindrical eye, is a second regularly shaped eye. Both pairs of eyes are fully functional and developed including retina and lens, providing vision unparalleled in the ocean. Indeed no other backboned animal, either on land or water, has developed four separate eyes. This is just one adaptation that living in such hazardous conditions can result in.

The barrel-eye spookfish also has binocular shaped eyes, although in this case no additional eyes. As its tuna shaped body clearly indicates this is a much more active hunter than its delicate relatives. It also travels to waters shallow enough that a silhouette is visible, revealing any prey above it but simultaneously revealing itself to any hunters beneath it. The barrel-eye has found a way of dissolving its silhouette. Its rectum harbours an organ filled with glowing bacteria. The resulting light is scattered across the fish's flattened underside enabling the barrel-eye to blend seamlessly with the pale waters above it. But like the pineapple fish it is unable to control the amount of luminescence and so has to travel vertically throughout the day in order for its camouflage to remain effective. If the barrel-eye gets distracted and delays on its daily migration, say in the process of catching a meal, than its shape will increasingly stand out to the keen eyed hunters rising from below.

Even the most advanced eyes still require some light to function. Below two hundred metres sunlight does not penetrate at all. But that does not mean it is pitch dark. As well as the lanternfish a multitude of other fish use and control bioluminescence both to hunt and communicate. One group in particular, the dragonfish, has used this ability to become one of the most diverse and numerous groups of hunters in the deep. The scaleless dragonfish offers a glimpse into how this group may have first started to use light.

It is far from an attractive creature; with its oversized tooth-laden head and dragging behind that a body that looks like it has just been run over by a lorry. Yet beneath its startling head is a glowing barbell filled with photophores, a tempting enough sight to entice potential prey to investigate further. This species has a small but expandable mouth, similar to a python, which allows it to lunge at and overpower large meals. Most dragonfish are more evident with multiple rows of photophores running along their bodies which may be arranged in spots or even long stripes. Other species emphasise the glowing barbell, extending to it such lengths that it can swing below the jaws while the rest of the body blends inconspicuously with the background.

The fangs lining the jaws of a dragonfish are nearly as formidable as those of the fangtooth, but they have an additional advantage; under water, which they are in their entire life, the teeth are transparent. Most of their victims won't get a chance to notice the teeth until they are gripping into their flesh. One particularly visually striking species, Sloane's viperfish, has teeth so massive that it is incapable of fully closing its mouth, to swallow its prey it has to physically dislocate its jaw- and fortunately for the fish they can cope with being opened to a ninety degree angle. Rather than having a glowing light *under* its chin the dorsal fin is elongated into this exceptionally flimsy filament that can stretch half the length of the fish and has a glowing light at its end. As if that wasn't enough rows of photophores run the length of the body; all leading back to those exceptional teeth.

The stoplight loosejaw is a significantly smaller dragonfish but has the most extreme jaws of any fish that can match even the viperfish when it comes to uncomfortable dislocation. To reduce the effects of drag while lunging for a meal its lower jaw has lost most of the typical features. It has no muscles, no protective sinew and even lacks any skin; enabling them to project rapidly and grasp a passing shrimp with minimum force. Indeed the supporting bone framework is virtually the only significant part of the lower jaw that still remains. Its teeth are tiny by viperfish standards, but they're not there to grip but instead to hook into the hapless crustacean while the jaws ricochet back towards the throat.

The loosejaw has another secret weapon which is equally unusual. Behind each eye is the standard green photophore which is easily seen by its prey but just in front of the green light, positioned immediately below the eye, is an additional photophore. This produces light which ranges from standard red to Infrared; well beyond the visual capacities of the human eye.

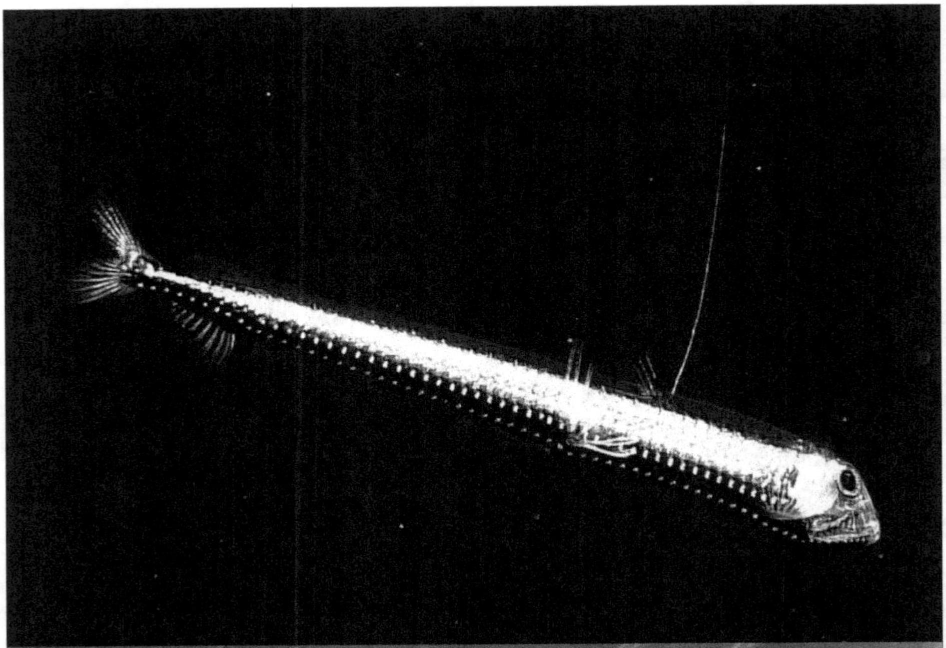

Danae Viperfish -*NOAA Ocean Exploration*
Sloane's viperfish, 870km North West of Puerto Rico- *Peter Rask Møller*

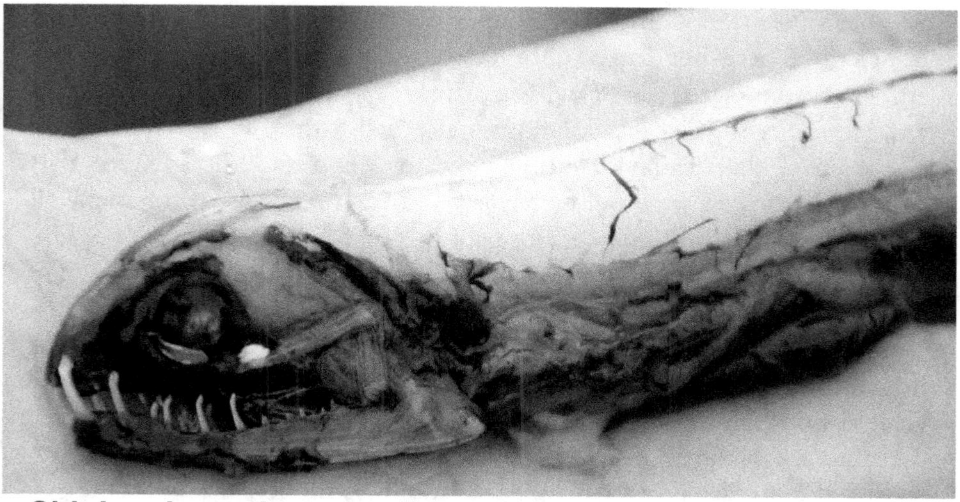

Shining Loosejaw, small enough to be held in a hand, Pacific Hake Survey, Washington- *Grant Law*

In the darkness few deepsea specialists are able to see either of the latter colours; except of course the loosejaw, and a handful of its relatives. As a result, they can illuminate and snap at prey while the potential victim is still moving closer to investigate the visible green light. The extra light's major limitation is that the colour red does not travel far under water, meaning that other fish can feel the vibrations resulting from the dragonfish's swimming and flee well before the predator can clearly see its potential victim. Consequently, rather than migrate vertically like many other marine hunters the loosejaws stay in the deep where the most common type of prey are tiny marine copepods (distant relatives of the crab and significantly more abundant in these waters than any type of fish). If any fish are unlucky enough, however, to stray near their mouth they will not hesitate to attempt to gobble them up also.

Such snap happy tactics cause a serious problem when it comes to mating. How does a dragonfish know if an approaching fish is a potential mate or just a particularly naive future meal? Each of the different dragonfish species has their own luminescent pattern as distinct as fingerprints from one another; some with great tassels hanging from their flanks, others with patterns of glowing skin under their chin. They also vary the timing of blinking their lights in the darkness. The female black dragonfish is one of the most impressive. At sea level she would be insignificant enough that a grouper could gobble her up and still feel hungry but here her forty centimetre length makes her a virtual giant among her kind- the loosejaw, for instance, would be lucky to reach 24 cm. Her body is a rich black colour, patterned with rows of blue and red photophores, all leading alluringly to her open mouth, as does that glowing barbell swaying below her jaws. She is not alone. Intriguingly all of these large displaying dragonfish throughout the deep ocean are female. So where are the males? But being so obvious is not the only way of life available to dragonfish.

A worm shaped dragonfish shares the same waters. Its eyes however are more like those of snails; poking up on long stalks. It may look bizarre but this is in fact an infant black dragonfish. When it is born the dragonfish's eyes are already positioned the same distance from the centre of its skull as they will be as an adult. As the females grow the eye stalks are slowly absorbed into the head, until they are non-existent. By the time she is that size she may reach half a metre or more. Males develop differently. Immediately after hatching they begin to search for a mate. Barely a few centimetres long they pale in comparison to the females. The males are so immature that they don't even have a digestive tract. They need to find a mate before they starve to death.

It is not without purpose that baby male dragonfish are born with such long eye stalks; in the dark it is their ability to spot a flashing female that will determine their success. The distinctive patterns are the only clue the males have to identify their own kind. When a male does reach a female he is so small that she may not even notice him mating with her. The effort alone is enough to kill him. The grim reality is that if the male was to develop further he would likely be taken as prey or compete with the female for already scarce food; it is only through his premature death that he is able to reproduce at all.

Although spectacularly coloured dragonfish are surprisingly small. The combination of high water pressure and lack of food tends to act against large fish. But there are still some predators capable of making short work of a dragonfish. Eels. Here they stalk the open waters with an audacity that few of their cousins would attempt on the reef. Occasionally green morays or congers venture down here but there are other more specialised species down here. Just as the dragonfish have adapted to the deep with their efficient eyes and photophores the eels have developed their mouths to similar extremes.

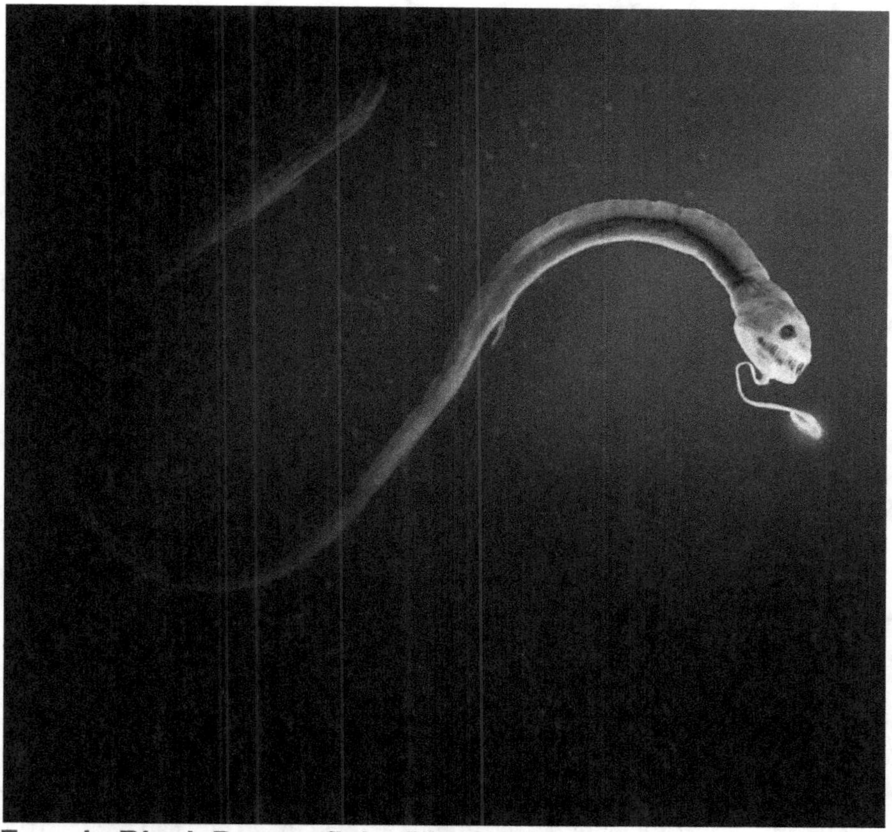

Female Black Dragonfish- *Diegograndi*

The sawtooth eel has a thin tapered beak that it can wield like a rapier against its prey. It positions itself vertically so that it can look for the silhouette of prey passing above. Most of the body is kept as rigid as a spear except for the very tip of its tail which remains flicking gently, preventing the eel from sinking. Only when content that there is no food in the area will it resume swimming in the typical eel fashion in search of better hunting grounds.

The similar shaped snipe eel also has a slim snout but its teeth interlace like a zipper. A shrimp snagged by its antenna against the teeth will be securely trapped while the eel's tongue slowly drags its meal into the throat. Indeed each tooth is locked in position with its counterpart until the catch has been dragged past them. But zippers have a tendency to snag and so do the peculiar teeth of the snipe eel. Under the pangs of starvation they might snap at a passing animal only to find their mouth caught on the tail of a fish well beyond their ability to swallow. The delicate snipe eel simply doesn't have the muscle power to pry open its own mouth. Nor does the fish it is gripping, in most cases, have the flexibility to turn around and pull the eel off its body. The snipe eel is left helplessly stuck clinging to its host, likely resulting in both of them starving or being eaten by something else due to their loss of speed. Despite this unfortunate fate enough of these peculiar eels survive to hunt and breed and maintain their numbers. Their backbones are as odd as their mouths for they have more of them than any other animal on the planet, up to 750. With such an impressive internal structure they can be more than a metre and a half in length, a virtual giant compared to its competitors in the deep. But their specialised feeding style means that it is not the snipe eel that is the ultimate master of its surroundings but another specimen from another, rather more formidable, family of eels.

The gulper eels have long bodies but for most of the time it is as slender and flexible as a skipping rope. It is only the skull that truly defines these animals as large. It is massively oversized; it looks like it belongs to another creature entirely. The head is huge but it is also a lesson in simplicity, with only its sky-blue eyes distinguishing it. The jaws are its most obvious feature and what jaws they are; stretching right back nearly to the base of the skull. The jaws look formidable enough when closed but it is only when the gulper gapes that one can get a true perspective of them. They are capable of stretching out to catch animals slightly larger than themselves. Once a catch is swallowed the stomach proves equally versatile; ballooning out, stretching the surrounding skin and muscle in the process.

The deep sea gulper needs to be able to detect its prey as well as it can swallow it. It has taken its sensitive lateral line to similar extremes as its jaws. Rather than lying neatly against the flanks the lateral line projects out beyond the body, unhindered by the protective layer of skin that other fish have. This provides more sensitive information about nearby movement; what's causing it, its size, and whether its potential prey or predatory. In shallower waters an exposed lateral line could easily be damaged by a rough piece of coral or other obstruction but in the depths such obstacles are rare. Even with the advanced lateral line there is still a poor chance of encountering suitable prey. So the deep sea gulper, like the dragonfish, has taken measures to attract a meal from a distance.

The tip of its tail swivels in a mesmerising display which is all the more palpable due to its illumination. A light organ gleams at its end like a jewel, but unlike any jewel it consists of flickering tentacles that continue to change the intensity of their colour. In the vast darkness any hungry fish might soon get tempted by this tantalising morsel. The deep sea gulper directs the fish like traffic, as they begin to nibble its light organ it gradually curls its tail towards its mouth. Then, in one swift contortionist movement, the gulper swallows its own tail; quickly sliding it back out of the mouth leaving the prey trapped inside. At times the gulper swallows a meal so large that it starts to rot before it can be properly digested. When this happens the rise of gases start to drag the eel up out of the depths where the change of temperature and pressure quickly cause it to perish. An unrelated group of fish (indeed they are more closely related to tuna than gulper eels) with even more flexible jaws have been found with such frequency floating at the surface with abdomens bulging in this way that they have been called deep sea swallowers; one particularly adventurous individual had a fish four times is own length within its gut. But the gulper eels themselves can not be quite as ambitious with their hunting abilities and some species have enlarged on even the deep sea gulper's formidable jaws.

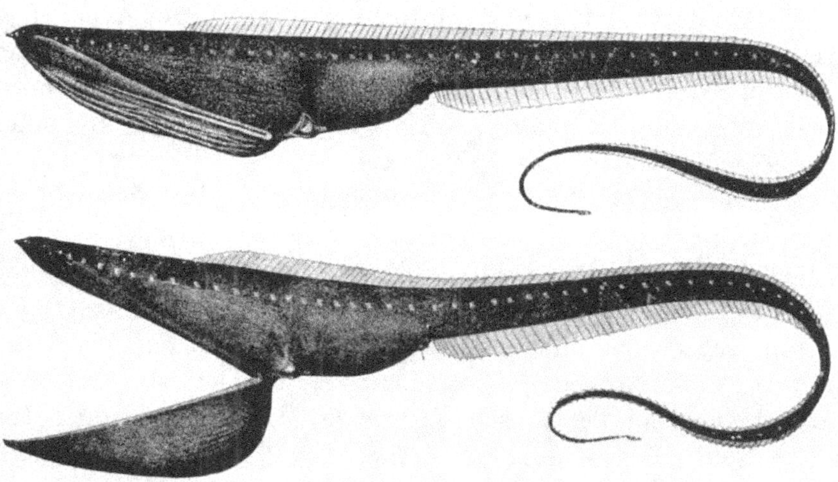

Elephant Eel- *"Oceanic Ichthyology" G. Brown Goode & Tarleton H. Bean, 1896.*

The most extreme mouth of all belongs to a close relative of the gulper eels; the pelican or elephant eel. Both names are equally apt. The elephant gulper is the largest, although not longest, of the deep sea eels, at seventy centimetres long it is more than twice the size of most of its relatives. The light organ on the tail is comparatively primitive but that only reflects the importance of its jaws. The mouth is not only highly extendible but has a pouch just like a pelican, hence its alternate name, allowing it to hold surplus food in the rare case that the gulper's stomach is too full to accommodate more. Despite its size the fish actually specialises in miniature prey. Shrimp occasionally shoal in the depths and with a great swoop of its throat the pelican gulper can easily swallow many of the struggling invertebrates in an action reminiscent of a whale. Even with its large mouth the gulper eel still has difficulty finding enough food to survive. This becomes most obvious when a pair comes together to mate. Producing eggs or sperm on such a meagre diet saps most of the eel's energy resources. Its bones begin to dissolve, releasing calcium into its blood stream. The muscles begin to deteriorate and the fish's body begins to finally fail. Eventually even its mighty jaws are completely dissolved and the eel is no longer able to breathe. Now the elephant gulper's sole means of survival is genetically, its abandoned offspring.

Despite their parents' investment juvenile gulpers are frail and have only a regularly sized mouth. In order to feed the youngsters have to hunt plankton individually, rather than en masse like their parents do. It will take them several months before they acquire the energy resources to devote to enlarging their jaws.

Sharks living at the same depth are just as adapt at dealing with the constant risk of starvation. Their lack of a swim bladder means that changes in water pressure are basically irrelevant to them allowing them to move with ease between different depths. Their gristle based skeletons means that they can also significantly stretch their jaws to overcome victims otherwise out of range. Like their coastal relatives electrical detection is a key feature in locating potential food. Hammerheads, which sometimes enter the deep, have an extended flattened head to increase their sensitivity. Goblin sharks also have a distorted head but in this case it is in the form of a horn like protrusion that sticks out of their skull. This horn, while slightly reducing their speed, increases their ability to locate both prey and mates. These are some of the largest deep sea fish, some surpassing three metres in length, but their passive lifestyle means they only have to eat every couple of months or so despite their girth. Goblin sharks live out their days without ever seeing the sun, but their other keen senses are all they need to navigate around. By contrast there are some deep sea bony fish with bodies and senses so de-conditioned that their existence is solely dependent on their partner.

Male deep sea anglerfish are insignificant physically but this is not the case when considering either their feeding or breeding techniques, which are unparalleled in the animal kingdom. These are distant cousins of the monkfish and frogfish but the pressure of depth has led to vast changes in their body plan. The black devil angler male is a tadpole shaped wriggler so underdeveloped that he is unable to feed and is unlikely to live for long. Not so the female who may be five times his size, at eighteen centimetres long, and can live for several years. She floats passively in the darkness, like a black balloon with a lantern shaped ray jutting from her forehead. The lantern is actually a specialised glowing organ, an esca, lit by millions of bacteria. The ray is waved constantly above the female's head, acting as a clear beacon both for confused small fish and the miniscule male. As he reaches the female he ignores her glowing esca and instead bites near her tail enabling him to keep up with his muscular mate. He remains clinging to her until she releases a stream of eggs. Once this is done the male black devil fertilizes them and swims off. He is unlikely to find another opportunity to mate again before he starves to death.

A male hairy anglerfish is also significantly dwarfed by his partner but their relationship is rather more intimate. A large white organ on his nose allows him to detect the specific chemical smells of a female, until he is close enough to visually recognise her. He still bites her but sharp scales around his head also hook into her skin. The linking of scales stimulates the increasing bond between the pair. Over several days the bloodstreams of the pair fuse and the male becomes fully supported by his larger partner. His mouth gradually changes position so that he can continue to breathe independently. He is unable to feed but all his dietary needs are already taken care of. Eventually he dwindles in size until he is little more than a living breathing, testicle; all other functions being taken over completely by the female. She even controls when he releases sperm so that their spawning can be precisely coordinated.

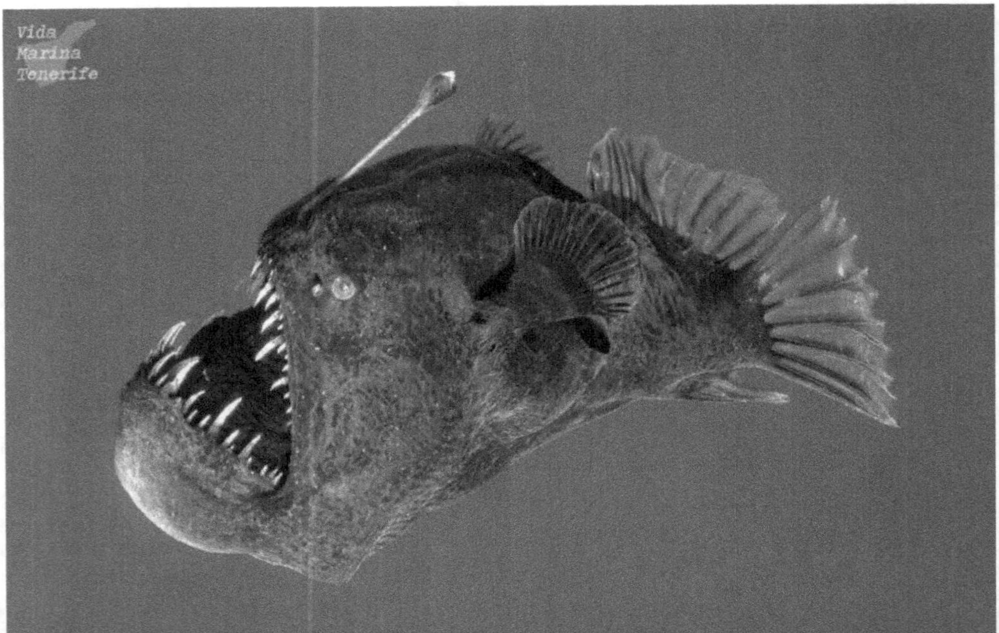
Vida
Marina
Tenerife

◄Female Black devil angler, Canary Islands- *Marc Martin Sola*
◄Juvenile whipnose angler, 200m depth off the west coast of Florida- *Gabriel Jensen*

The male deep sea angler, despite appearing little more than a convenient parasite, also has significant effects on the female. For many years it was not understood why deep sea anglerfish females, alone of all fish, only become sexually mature after mating rather than on reaching a certain size like other fish. But the new research on the interrelationship between gills and breeding sheds new light on this. Supporting a male reduces the effectiveness of the anglerfish's gills and it is the resulting drop in oxygen that, as in all other fish, likely causes her to finally reach maturity. Those females that fail to attract a mate never develop into adults and will remain in perpetual adolescence until a male becomes attached. But with such visually demanding displays even in the depths of the ocean such an event is unlikely.

Occasionally a female will play host to several males, all clinging inertly to her in such a way that when they were first observed scientists understandably misidentified the attached males as juveniles and assumed the mother was feeding them; you could say they were half-right.

The hairy anglerfish attract preys to her just as she does a mate; using irresistable light display. But she also uses something else. Emerging from her body are multiple slender rods, rising like antennas; enlarging the area that the anglerfish is able to monitor for food. As well as bearing a sensory purpose these rods also act similar to the gingerbread house of fairy tales, entrancing naïve passers-by close enough to be consumed. It is only when one of the antennas are triggered that the hairy angler demonstrates her extreme flexibility, turning in a flash to consume whatever was foolish enough to disturb her. She can even lower her lure so that the glowing esca is dangling directly in front of her lips. On some occasions she may even inadvertently sever and swallow her own lure along with her meal but fortunately she can soon regrow it.

The whipnose anglerfish group have no such additional features, instead each species focusing all of their resources into upgrading the lure. For many species in this family the esca itself resembles an archaic oil lamp, with its supporting network of flimsy struts. The lure is thrust vigorously around the fish, indeed in some species the rod can be three times the size of its owner. This would be curious enough by itself but many fish display even more elaborate behaviour. Elsman's whipnose swims upside down, wriggling its rod above the mud of the sea floor, drawing out hiding invertebrates. The wonderfish has her lure in an even more remarkable position. It actually hangs inside her throat, attached to the roof of the mouth. Any prey tempted to investigate are in the ideal position to be swallowed- head first. Having the lure deep inside her throat also reduces the chances of her inadvertently biting it when hunting.

Each of the different deep sea anglerfish has its own way to use its lure and hunt and it is these differences which also allow each species to be easily distinguished by people, and likely each other. Part of the group's success, it must be admitted, is due not to the anglerfish themselves but to the colonisation of their lures by luminescent bacteria. Despite the hundreds of specimens that have been dragged up from the deep it is still not known exactly how the juvenile female anglers manage to pick up their essential crop. Yet somehow there is one deep sea angler that manages to survive without its bacteria and uses its body more bizarrely and effectively than any of its relatives.

The hairy seadevil looks like a crude cross between a fangtooth and a hairy angler but its most significant feature is not its lure but something rather different. A pair of brush like fins growing from its abdomen and spine combs the surrounding area, snagging any small invertebrates that tangle in them. Once the brushes are triggered the female twists them towards her and picks them clean. Other outgrowths jut out like skewers across the forehead, increasing in size the closer they are to the lure. And it is the lure which is the real surprise feature of this hunter. The lure is overshadowed by such a dense layer of modified scales and fins that it can barely be perceived, it certainly does not glow. Indeed the lure does not release any light at all due to the simple fact that alone of all its kind the hairy seadevil is independent of glowing bacteria. The hairy seadevil is left to feel its way around until it blunders into something worth its time. In such exposed depths it would be thought that there would be nothing to guide its position but there is one immediate barrier perceptible, even to the seadevil, the ocean floor. There is food to be found here too but most of it is rotting and immobile and consequently unlikely to be discovered even by this remarkable anglerfish.

Living permanently on the ocean floor means saving any energy possible, and one fish personifies this more than any other. The tripodfish rests on the sea floor on three long fins. It stays motionless facing into the current, waiting for any potential morsel to brush past. And it can wait several weeks between meals, due to the considerable effect the cold water has on slowing digestion. With its unusual fins the tripod fish rarely swims. But it doesn't have to. It can not only rest on its specially elongated fins but walk with them, striding like a curious insect. While its gait is peculiar it is still much more energy friendly than active swimming.

For the most part there is so little food here that a tripodfish is unlikely to see another of its kind throughout its life. In other fish this would cause problems with reproducing but this solitary creature has a more than adequate solution. Many ray finned fish are sequential hermaphrodites, able to change from one sex to the other, but the tripod fish has the capacity for both its male and female sex organs to be active at once. Other fish like the bonnethead shark that reproduce without mating give birth to exact clones but in tripodfish both organs produce slightly different genetic material meaning that the offspring will inherit at least some differences from its parent. This increased genetic diversity will aid their survival under such difficult conditions; enabling them to endure slightly different conditions from their immediate relatives and potentially increase their chances of survival.

Tripodfish, Honduras- ***Benoît Segerer***

The tripodfish lives its entire life without ever seeing the sun rise, a spring bloom of plankton or the famine of winter. At these depths the ocean also maintains a constant frigid temperature year round, about four degrees Celsius above freezing. But surprisingly enough the seasons do still make their mark here. As vast shoals of sardines and the like swarm kilometres above them inconsequential scales flake away and begin to sink unnoticed towards the seafloor. Occasionally whole corpses will follow their skin cells into the darkness. By the time the carcasses reach the seafloor it is well into Autumn on the surface but down here it may as well be Spring. This is the time when many of the local inhabitants come together to breed. Tripodfish are fairly restricted in this but more athletic fish do so in spectacular numbers.

The orange roughy is as capable a swimmer as can be expected in the depths and this allows them to travel for miles to breeding territories. Millions of them gather around, announcing themselves with strange auditory clicks produced by their swim bladder. Their numbers can only be truly appreciated by the tremendous catches of this species; in the peak of their harvesting during the middle of the Twentieth Century fifty million kilograms were trawled annually from New Zealand and Australia, and this from a fish where only the oldest get close to five kilos.

Such a large population of fish usually indicates animals that breed and die young; producing thousands if not more offspring in the process. Orange roughy are different. The age of a bony fish can most accurately be worked out by counting the amount of layers in the otoliths, or ear bones, which typically increase annually. And the sheer number of layers in the roughies' otoliths is beyond expectations. Their restricted food supply and low temperatures has had the response of slowing down the roughies' pace of life. Many of the fish examined were at least a hundred years old. Others reached one hundred and fifty but even these were still comparatively middle aged. A few rare otoliths revealed an almost unbelievable age of the, now inevitably deceased, fish which contributed to the research; individuals that had reached two hundred years of age. As these were not fish that had naturally died of old age but had instead been caught by either net or trawler it is unknown what the exact limit of the roughy's longevity is. The orange roughy is among the longest living of all bony fish and only the slovenly Greenland sharks and the land tortoises surpasses their longevity among other vertebrates.

Living for two hundred years comes at a cost; a sluggish reproductive rate. An orange roughy is not sexually capable until is about thirty. Even then there is no guarantee that it will be able to find a suitable mate for several years. Any significant change in habitat or rate of predation could soon cause extinction of the whole population. Indeed, in some areas these roughies have already been overfished or had their habitat so ravaged by trawling that they are unable to find sufficient food. Certainly catches of orange roughies in the millions of kilograms no longer occur, nor are legally permitted.

Even at their most numerous the roughies were not the most successful of deep sea fish. That title belongs to another family related to the cod and anglerfish with the uninviting name of rattails or grenadiers. There are between three hundred and four hundred different species, each neatly suited to its own particular niche in the deep sea and many of them similar enough to have confused scientists about their identify. Some are barely ten centimetres long while others can be more than a metre. As juveniles their diet is mainly slovenly invertebrates near the coast, like sea cucumbers or crabs, but as adults they descend, targeting fish and swift creatures like squid. Their diet requires them to be streamlined and fast; driving themselves with the aid of that awkward rodent like tail. Food is still scarce so they only use their speed when hunting, otherwise they flap their fins just enough to maintain their position. Even so the grenadier is still active enough that it has to feed regularly. Grenadiers have modified hearing, acoustic sensors which allow them to detect and track the gentle patting sound produced by fin strokes. If it is unable to hear or spy an immediate meal it digs its snout into the sand, using sensitive barbells to aid its search.

To camouflage themselves some grenadiers have photophores glowing from their rectum, thus reducing their silhouette from any creature hiding in the substrate below them. If they are lucky they might encounter an easy meal, such as a half eaten dolphin drifting down. A large carcass is a bonanza and several grenadiers arrive to feast while it still remains. Such meals provide a rare opportunity for mating. While they are unable to see each other they can certainly hear each other thrashing around and males have developed ways to stand out further. They have large muscles which they beat against their swim bladder, producing a drumming sound. Males spar noisily beside each other until the weaker individuals fall silent.

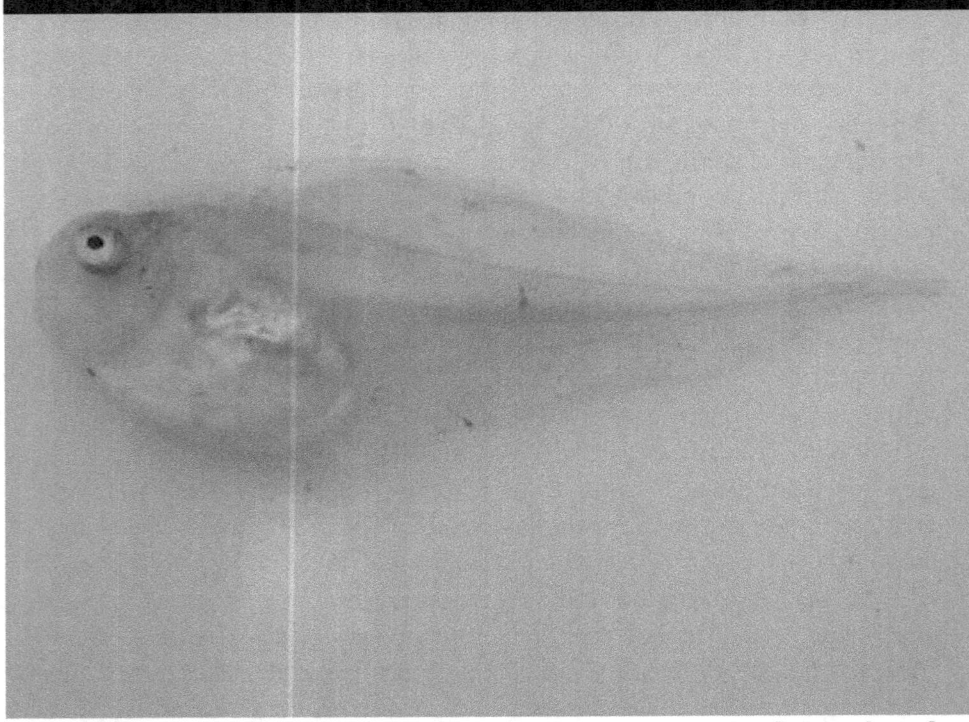

Rough Abyssal Grenadier- ***U.S. National Oceanic & Atmospheric Administration***
Careproctus Snailfish- ***Erica Acuna- Sea Survey***

Grenadiers are so capably adapted to the deep that they descend far beyond the range of anglerfish, gulper eels and even the stalwartly lantern fish.Many reach three thousands metres into the deep and one species lives as far as six kilometres beyond the sun's rays. A few specialised fish live even deeper, under some of the most extreme living conditions possible.

As the depth increases the pressure on the fish becomes more severe. On land a similar effect can be seen in the change in altitude that occurs as you ascend a mountain. The highest mountain on earth, Everest, rises 8,850 metres above sea level and its peak is uninhabitable even by that most adaptable of species, humans. The few animals that can survive on some of the mildly lower peaks are specialist survivors. Their bodies for the most part are squat, keeping them well insulated from the pervading cold.

Species living at these depths are also relatively leisurely in movement compared to their lower altitude relatives, allowing them to save precious energy. So it is not surprising that in the depths of the ocean the difference in living conditions also changes fish significantly from their relatives at more hospitable levels. For crisscrossing the world is a series of oceanic rifts that in places could submerge the Himalayas completely. Here the separation of the continental plates has ripped great gashes in the oceanic floor for thousands of kilometres. They are the largest geographical feature on earth, with the exception of the continents alone, but completely obscured by the world's oceans.

The most significant of these great gashes is found in the Pacific Ocean between the coastlines of North America and Asia, the Marianas Trench. In one special part of this oceanic trench the ground simply drops away, for more than 11,000 metres below the surface. This is the most inaccessible place on earth and has been called, rather appropriately, Challenger's Deep. The pressure here crushes any solid object to pieces. The majority of this great abyss is known only through its detection with radar. Few submersibles have been built that can survive the journey. Man may have walked on the moon but until recently had only reached the depths of Challenger's Deep twice, once in 1960 and again in 2012. Since 2019 technology has advanced enough for a submersible to commence regularly taking tourists deep into the trench but even with these significant technological advancements limits still exist. No human is yet to venture outside of the submersible into the deepest ravine in our planet; nor is this likely due to the tremendous outside pressure. Consequently the sea here has been left in sole possession of a completely unknown community of fish and invertebrates. So little known are these great depths that we are completely oblivious as to what fish exist here or if, like the peaks of Mount Everest, such an area is largely uninhabited.

In these depths, discoveries are frequently made that overturn established record holders, and few records are more impressive than that of the world's deepest fish. In 2022 a new record holder was uncovered in the Izu-Ogasawara Trench off the coast of Japan. It was a snailfish. Snailfish are not exactly new to science, there are more than 400 species of these translucent frog-like species, but this one *was* new. Its pink gelatinous body stands out as does its toadlike head and its oversized pectoral fins that fan out over the body like some fancy-dress costume. So does the depth it was found; 8,336 metres below sea level. Its gelatinous body is part of its secret; the high layer of mucus covering it reduces the level of friction but it is also easier for the body to make than standard skin, both factors significantly reducing its energy requirements. The fish discovered wasn't on it own either; there were several of them, all juveniles. Their youth is part of their ability to survive here, adults are less able to cope with extreme water pressure, and even the observed juveniles were close to the very limits- although that very same pressure also prevent most predators from entering the same waters. Scientists have calculated the maximum depth of pressure after which a fish's body would be unable to function, particularly the manufacture of proteins that many fish (including snailfish) use to allow their body to cope under high pressure, and came up with the figure of 8,400m.

Even for snailfish that is extreme and many of the group's species never venture that deep even as youngsters. One group of snailfish, the Careproctus, have found other ways to keep their young as safe as possible. In at least one species the father keeps the eggs in his mouth until they are ready to hatch, like the arowana does, but other Careproctus species resort to babysitters. They target the most unlikely of potential parents, king crabs, laying their eggs upon the gill cavity of the crustacean. The eggs readily stick and clump together, being constantly ventilated every time the crab attempts to breath. Some unlucky king crabs get targeted by so many snailfish that their gills end up being blocked completely and they soon expire- along with the attached eggs. At least one Careproctus species has an adaptation even for that its developing young have a lower energy requirement than most other marine fish, resulting in them having reduced oxygen requirements, and compete less with their unwilling host as result. Even after hatching the snailfish mightn't necessarily be done with the king-crab and may attach itself to the crab's back or even its legs, allowing itself to be transported, practically energy-free, to potential new habitats.

The Puerto Rico Trench is home to the other potential contender of the deepest fish in the world, the deepsea brotulas. Strange, pink and wormlike they would not look out of place in an alien movie. Their odd shape is a direct result of the unusual conditions forced upon them by this extreme environment. The deepsea brotula is indeed adventurous in the depths it will travel to. They regularly live in water several kilometres below the surface. But they can go much deeper than that.

The deepest brotula found so far was at a depth of 8,370 metres, even deeper than the snailfish, but the fact that this individual was discovered in an open trawl net means that it could have been potentially been trapped in shallower waters before the net was lowered. In the inaccessible darkness there are large gaps in our knowledge of the brotula and other deep sea species. But relatives of the deepsea brotula living in slightly shallower waters can shed some light on their natural behaviour.

The Galapagos Islands are internationally recognised for their unique terrestrial wildlife but two kilometres beneath the coastline is an equally diverse, and if anything even more extraordinary, ecosystem. Most of the water is freezing but in a few select areas submerged volcanoes belch up great clouds of toxic hydrogen sulphide. These gas clouds are significantly hotter than boiling temperature; 450°C. This is far beyond the capabilities of any fish but the water flowing around these hydrothermal vents is temperate enough to be able to support a surprising amount of life.

Giant worms, three metres high with a protective layer across their body, filter the gas out of the boiling water using specialist bacteria to transform the chemical into energy. They also inadvertently shield other life from the direct heat of the volcanoes. Nibbling on the worms are crabs and other small invertebrates. And darting after them all is a bright pink relative of the deepsea brotula, the pink ventfish or eelpout.

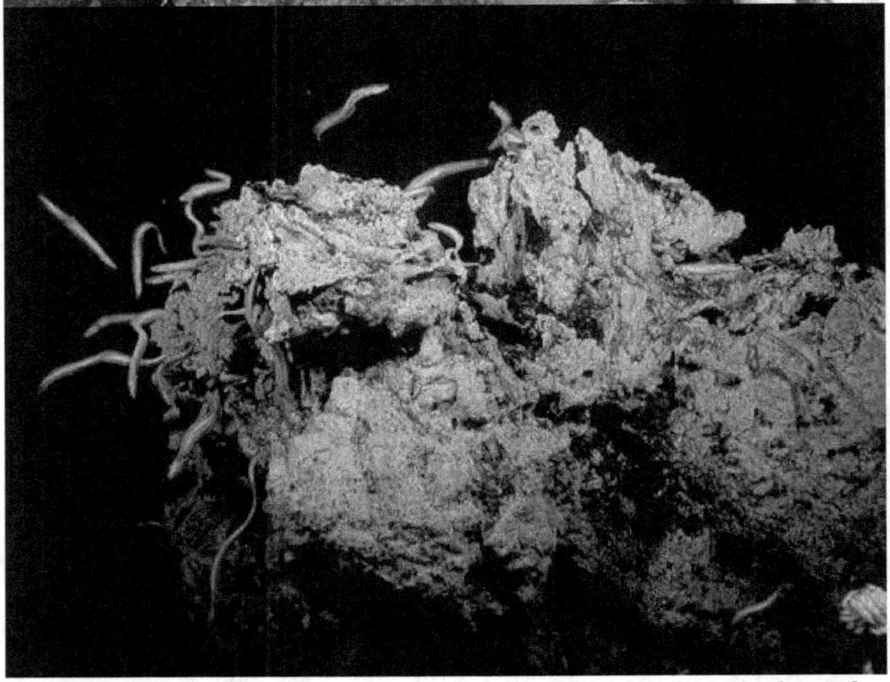

Pink Ventfish, East Pacific Rise vent fields, South America- ***Woods Hole Oceanographic Institution***
Cutthroat eels, Eel City Vailulu'u volcano, American Samoa- ***Vailulu'u 2005 Expedition NOAA-OE***

Juvenile Trevally, Aquarium in Dubai Mall- *Tan4ikk*

This is the key predator of these deep sea vents; they will even turn on the massive worms sheltering them from the boiling water in order to have an easy meal.

But the ventfish doesn't attack all small invertebrates indiscriminately. Crabs for instance can provide more than just a nutritional service. They also provide valuable hygiene service. As the ventfish rests a white crab crawls over to it and begins nipping off small parasites. The pink fish remains practically motionless throughout the procedure, just as groupers and moray eels allow wrasse to clean them on the reef. Once satisfied the ventfish dashes away to snatch up any unsuspecting prey.

While the ventfish can cope under such perilous conditions any of its eggs laid here would be at risk of drifting into the boiling hydrothermal vents and inevitably perish. So the ventfish, like many of its relatives, gives birth to highly developed live young that can negotiate their way around their dangerous home almost immediately. Volcanic currents are surprisingly fickle however and over a decade they may shift course, destroying the whole ecosystem as rapidly as an out of control forest fire can wipe out a rainforest. Just as the current may disappear so in other areas will they suddenly appear and the drifting eggs and larvae of the worms will take hold allowing another ecosystem to be formed. It will take, though, until enough worms are established that the ventfish can find food there. For the lucky few the occasional sunken whale carcass may provide enough sustenance to last them until they reach the next deepsea vent. Some ventfish will be able to barge their way into other territories, but those unable to do so must either starve or are forced to hunt among the plankton.

Across the tropical waters of the Western Atlantic, the Carribean and Eastern Pacific a specialised eel, a member of the cutthroat eel family, also lives around the hydrothermal vents. This species swarms in vast numbers; around the vents of Vailulu'u volcano in American Samoa they are so numerous that it has resulted in their favoured location being nicknamed the 'Eel City'. These eels have an energy level unmatched by many other fish at these depths, moving at speed through the super-heated waters. Unlike the pink ventfish they are unable to find enough sustenance in this habitat to survive so to feed they leave the comparatively balmy conditions and venture into the near freezing open ocean feasting on plankton, particularly that of crustaceans, returning to the sanctuary of the vents whenever they need to recover.

Some fish spend their entire life tracking and living among the plankton. Indeed much of the plankton is made up of juvenile fish as well as invertebrates and eggs. The eggs of many deep dea fish are red, rendering them as invisible as their parents. But after hatching juveniles have to find other ways to protect themselves.

Infant trevallies and jacks play a dangerous game sheltering around the deadly tentacles of jellyfish until they are large enough to either outpace or consume their original pursuers. Jellyfish are far from the most benevolent of hosts but few hunters would try to swallow one to reach the sheltering fish.

A juvenile ribbonfish swimming seemingly unprotected in the open would also be a hard act to swallow. Depending on the species the fins jutting out from the fish can either be trailing filaments or slender spears. The infant is tiny but by spreading most of its body size across long fins it makes its small size count for all its worth. Adult ribbonfish have even more effective ways of dretracting predators. They lack awkward fin ornamentation which allows them to rely on their speed to chase down squid and escape most predators. But dolphins and several sharks can still overtake them. So the ribbonfish produces a defence usually only seen in squid. Near the tail is a large sack filled with dense dark liquid. But this is no buoyancy aid. When threatened the ribbonfish squirts this liquid behind it, blackening the water with an ink like substance just like an octopus, and rapidly makes its escape.

One species of ribbonfish, the oarfish, is the longest of these open water bony fish. It glides deep under the water plucking off individual prey in such a way that it is barely caught by either fisherman or scientist and hence little known.

Most oarfish sightings are after a severe storm when they risk stranding on shore. The largest of those specimens that has been reliably measured was eleven metres. Out at sea many of the oarfish seen, and there have not been many, appear to well outstrip the size of those beached specimens. Some of the field estimations of the oarfish's length are more than seventeen metres long, although it must be said such estimations are hard to verify. They might not weigh as much as a whale shark but if the measurements add up then some individuals could be the longest fish in the ocean. The nickname for the giant oarfish is the king of the herring and even though it is unrelated to herring the frill across its neck and its size certainly make it appear regal enough.

Some researchers have tried to link the oarfish with medieval sightings and depictions of giant serpents and although it does bear a striking resemblance it is unlikely that the fish ever came close enough to the surface to be seen from a passing ship. Perhaps a few beached oarfish might have been enough to inspire the legends; as recently as June 2025 the strandings in the same week of two oarfish in New Zealand, and a third in Tasmania, were considered apocalyptic warnings of earthquakes. The oarfish had been sighted prior to the 2011 major earthquake in Japan, a country where their traditional name translates into 'doomsday fish'.

However, by the time of the more recent strandings the scientists had already demonstrated, six years before, that the oarfishs' assumed association with earthquakes, or other adverse weather events, was inaccurate. To come to this conclusion the scientists had scoured the historical records, particulary in Japan where the original folklore that inspired this belief is from, for oarfish and other ribbonfish sightings and compared them to the dates of tsunami and earthquake events.

Some creatures are able to detect, and behave differently during extreme weather events; the most notable being elephants picking up earthquake vibrations through their sensitive feet, and then moving to more stable locations. But the data isn't showing that for the oarfish.

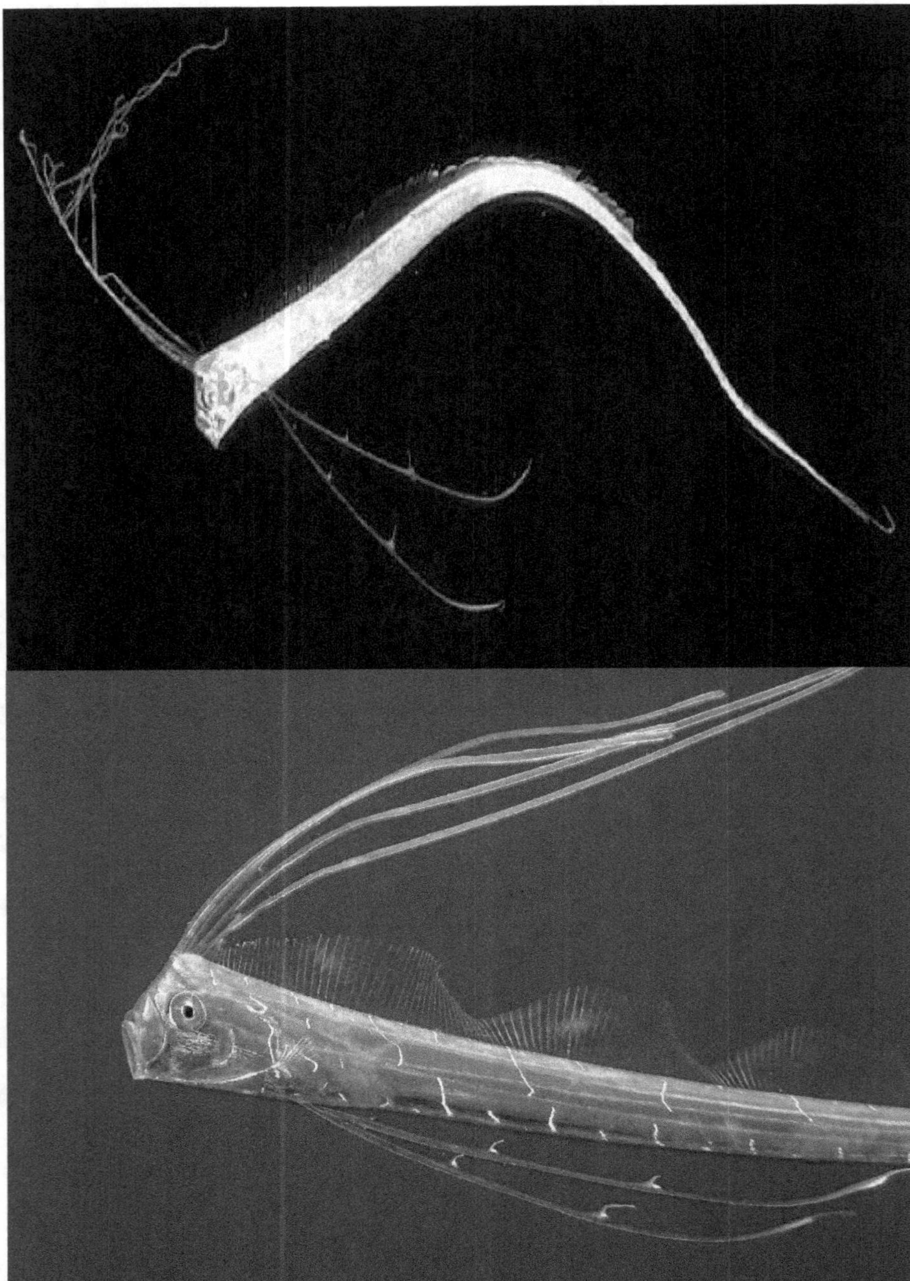

Juvenile Oarfish- ***NMFS-PIRO Observer Program***
Close up of juvenile Giant Oarfish- Santa Cruz, Spain- ***Marc Martin Sola***

There was no link between the fish appearances and the earthquakes or tsunamis. Their sightings occurs so irregularly though that this isn't a legend that is likely to disappear anytime soon.

The oarfish itself is no less curious than the myths it has inspired. The fish's scales contain high levels of ganoin, a silver covered mineral that reflects any passing light. Projecting from its head is a crimson crest with several fine filaments trailing behind. This crest merges with the dorsal fin which is the same vivid colour and continues all the way to the tail. Oarfish have no tail fin, their body simply trails to a point. Having such a long body complicates their movement so the oarfish has developed a swimming style all of its own. From the few observations of live fish it tends to position its body vertically, relying on its dorsal fin to wriggle its way around. Its two pelvic fins are notable not just for their length but also for consisting of only a single ray each, making them resemble oars. But rather than being used as oars they are held out to stabilise the fish as it waddles after plankton.

Other close relatives of the ribbonfish have sacrificed length for speed. The opah darts through the water after squid; at a pace its longer cousins could never achieve. The opah's mouth is toothless but also huge; allowing it to target larger food than just plankton. Its body is concentrated allowing it to move rapidly after prey but like the oarfish it doesn't rely on its tail to swim; instead it uses its pectoral fins to propel itself. Their speed is swift enough, and food in enough abundance, that they can grow to more than seventy kilograms. But there are even heavier bony fish in the oceans; giants that are measured not in kilograms but in tonnes. Their size restricts them to hunting slow prey, mainly plankton. But not all plankton is microscopic. Several jellyfish, for instance, stretch for several metres across. They are also armed with deadly poisons, long lines of tentacles tipped with neurotoxins able to be fired into any intruder. Due to their defences jellyfish can afford to lie just under the surface of the water, gleaming in the sunlight. This attracts their most formidable predator and the first indication you may have of them is the sound of them slurping up jellies as their mouths jut out above the waves.

The creature's massive body resembles a balloon that has expanded to such a circumference that it appears about to bust. The fish is about four metres across and weighs two tonnes and three hundred kilos. It is hardly surprising that when scientists were originally attributing names to animals they gave it the Latin name for an object equally as large, the millstone, *Mola mola*. Other people encountering it noted a resemblance to the celestial bodies orbiting above them and gave the fish its common name, the oceanic sunfish, even if its silvery grey complexion is actually more reminiscent of the moon.

Recently, and rather unexpectedly, an even larger species has been identified, a close relative- the giant sunfish, one monstrous individual caught in Portugal was more than two tonnes and seven hundred kilos. Both species, and a third significantly smaller cousin, all share the same signature shape. Rising above the curve of the body is a single enormous dorsal fin. To keep the sunfish stabilised an equally proportioned anal fin projects beneath it. Both oversized fins look stiff but they can be flexed to provide steering while its miniature pectoral fins provide most of the swimming motion. The tail fin and pelvic fins have disappeared altogether. Like our coccyx the original supporting bones for the tail have been fused together, resulting in an untidy shaped frill at the base of the fish.

Oceanic sunfish, Magadalena bay, Baja California, Mexico- ***Image-Source***

Oceanic sunfish feed mainly in the frigid waters of the depths, where jellyfish spend most of the day, so it is important that they warm up regularly beneath the sun before they submerge again. This also allows other creatures to take advantage of the parasites growing on the sunfish's skin and snatch an easy meal. Sunfish will even approach wading seagulls and tilt to one side allowing the bird to target the worst problem areas. There are other fish that can also appreciate the smorgasbord to be found on these moving behemoths. Remoras basically survive living on the backs of sharks, whales and large fish; scraping molluscs and crustaceans off their bodies. A modified dorsal fin forms a disk above its head, allowing the remora to stick to its host so that it can maintain contact despite the high speed its companion may be travelling at. They will even cling to passing vessels and begin cleaning them the same way. But staying attached to oceanic sunfish comes with a particular problem. During the larger fish's daily search for food they enter water cold enough that any attached remora would be frozen to its death. Nor would they be likely to survive long on their own if they abandoned their host temporarily. They desperately need a place to hide and such places are rare in the miles of apparently empty open ocean.

As the oceanic sunfish begins to decend the remoras dart into the only shelter in sight, beneath their host's gill cover. The large fish's gill cover is easily big enough to contain the remora and allows enough airflow for the giant to still breathe. It provides convenient insulation for the remora from the frozen depths that the gentle giant is about to enter. The rotund shape of the sunfish means that the behemoth maintains a stable temperature long enough for it to feast on jellyfish before rising once again to warm up in the sun, where the remoras can resume attending to its personal hygiene.

Other relatives of the remora are much more active inhabitants of the open waters. The cobia or black kingfish has two major physical differences from the remora. Firstly rather than having an adhesive disk on top of its head it has just a few short spines separate from its dorsal fin. Secondly its size; it can be two metres long and weigh seventy kilograms. It is also a great traveller, but rather than hitchhiking it has to swim for itself, relying on its dense pectoral fins. Such a powerful swimmer is the cobia that it is found in all tropical seas except for Central and Eastern Pacific. With their great gaping mouths they plough easily through shoals of shrimp and other crustaceans, occasionally snatching up a passing fish.

A pair of Great Barracudas, Bali- ***Aarrows***
Remoras attached to nurse shark- ***Image-Source***

The remora and cobia have another relative which is an even more active hunter. It is so well adapted to its lifestyle that adults bear minimal resemblance to its relatives. The larva of the dorado clearly reveals its kinship as it is almost identical to a juvenile cobia. The developing larva transforms from a thin weedy specimen into a muscle packed torpedo shaped speedster. The head of the dorado flattens back like the bonnet of a car; creating minimal resistance when moving at speed. They are one of the most colourful hunters of the open sea with their mauve dorsal fins, green bodies and the blue marks splattered across their body. The dorado can also change its colour; in the heat of attack, or when caught by a rod, it flashes between such an array of different tones that it is aptly called the rainbow of the sea. Unlike the cobia its main diet is not idle crustaceans but other swift fish. It is likely that the mesmerising change of appearance confuses prey like flying fish enabling them to be caught easier. Due to their energetic behaviour dorado are called 'dolphins' in several countries and just like the cetacean the dorado often hunt in large packs to pin down prey. When thirty or so gather together the image of them rushing beside each other, flashing brazenly as they attack, is almost hypnotic. Few shoals of small fish can hope to survive these predators for long.

Barracuda on the other hand are anything but gaudy. These silver hunters stalk the open oceans relentlessly until they encounter food. While still muscular these predators are designed less like torpedoes and more like spears. Just as a spear is tipped with a sharp point so the barracuda's mouth is lined with a fearsome array of protruding teeth. Most of the time they are slow but steady swimmers when compared to dorado. Barracuda are curious, always approaching to investigate a potential food source, and it is not unknown for them to snap at humans. Indeed in the murky waters of the West Indies, where it is harder for fish to distinguish prey, over thirty attacks have been reported, most perpetrated by the great barracuda. Barracuda typically hunts small fish during dusk and dawn when most fish are disorientated and vulnerable to attack; for a hunter in a hurry snack-sized fingers and toes are too readily mistaken for an easy meal.

Near a reef a great barracuda can be nearly two metres long and can easily pin most prey on their own against the coral. But in the open ocean, just as lantern fish have nowhere to hide, the barracuda have nowhere to pin their prey. To catch a meal they must overtake it. Ocean dwelling barracuda are half the size of their coastal relatives and have more difficulty subduing and killing their food. So small species like the chevron barracuda often hunt together. They begin by encircling a small shoal of sardines, then gradually draw nearer until they are only a few metres away from them. Only now do they reveal an entirely different nature. They dart out and snatch a fish from the shoal at nearly fifty kilometres an hour. The sardine's spine is quickly severed by the teeth before it is swallowed. The barracuda rushes out to grab another sardine, impaling and consuming it just as quickly. But such speed cannot be maintained for long by the barracuda before it has to return to a more relaxed swimming pace. This explains the barracuda's different body shape to the dorado; it is not designed for constantly travelling at speed but instead as an aquatic sprinter. Having a quick turn of speed is essential for barracuda for other hunters tend to view *them* as prey. Shortfin Mako, slighter relatives of the great white sharks, patrol the oceans. Like other mackerel sharks they are warm blooded, due in part to their insulating layers of solid muscle. They are the fastest sharks; their cruising speed is 50 kilometres an hour but when they have to they can reach 75 kilometres an hour. At this pace they can leap nearly nine metres out of the water, far beyond the capacity of any flying fish, and high enough to catch a seabird on rare occasions. Some estimates of the shortfin mako's upper speed are as high as 100 kilometres an hour but in the ocean any measurement of speed is little more than a crude estimate. What is undebatable is that they have equally competitive prey.

Tuna shoal in massive numbers across the temperate oceans. Two species in particular, the bluefin and yellowfin tuna, have a high impact both commercially and ecologically. They tend to travel in large schools that occasionally join together. Yellowfin are the smaller of the two, at 2.5 metres long, and can be further distinguished by the distinctive colour of their pectoral fins. They have also been seen to flash their bright colours at each other during spawning. Yellowfin tend to rise from the depths of the ocean and gather towards silhouettes which tend to indicate prey but could also be due to a shoal of dolphins. Yellowfin prefer to hunt with dolphins, tracking down suitable prey together. Dolphins have to constantly return to the surface to breathe but tuna have no such concern and can prevent any prey's escape.

Tuna have an unexpected similarity to dolphins, one that is not visible to the naked eye. Every movement of a muscle, whether it's the stroke of a fin or the pumping of the heart, produces heat. In most fish this heat is quickly absorbed by the surrounding water. But tuna and their relatives are insulated in a layer of fat and muscle so dense that they have a higher internal body temperature than the surrounding ocean. Like dolphins they are warm blooded, although they have come across such a development independently. Their temperature control allows tuna like the yellowfin and the albacore to continue being active at night, when their competitors reduce their activity. In the albacore this extra energy is used to fuel the longest migration of any fish; 15,800 kilometres from one side of the Pacific to the other in the pursuit of food.

The bluefin tuna has taken these adaptations even further. It is so huge and so effectively covered in fat and muscle that it is the only tuna capable of staying in temperate waters year round. Like the yellowfin they are also attracted to the sounds of cetaceans searching for prey near the surface. Most fish have to beware of dolphins. But the bluefin is twice the size of most dolphin species. The largest bluefin ever caught was 4.6 metres long and more than 680 kilograms. During the hunt it could easily swallow a juvenile bluefin less than half a metre long but this rarely occurs due to the fact that tuna tend to form groups with other individuals of the same size, regardless of the species. The bluefin has a particular need to reach such tremendous sizes. When winter approaches other tuna species have to leave the temperate waters for the warmer but comparatively barren tropical oceans. Bluefin can remain feeding unchallenged in the cooler productive waters but only if they've put on enough fat during the summer. Bluefin can live for more than two decades, about twice that of their relatives, providing they do not fall victim to a predator.

But being perpetually fast and warm comes at a consequence. The tuna's gill covers are fused in position which makes it easier for them to maintain high speeds but also means that it cannot gulp extra water to aid breathing. Likewise if it stops swimming or is forced to slow down significantly it will suffocate. Its gills are among the largest of any bony fish to compensate for this but even when travelling at top speed breathing is still limited. For male tuna the only solution is simply to stop growing; resulting in them barely reaching more than two metres. But females are able to take advantage of a quirk in their anatomy to continue growing.

During the winter breeding season all mature bluefin tunas migrate to warmer waters. Each of the three subspecies converges on a single breeding site. Thousands of tuna squirm together and the females begin to produce their eggs. Ten million modest sized eggs are produced, totally out of proportion with the size of the fish.

Yellowfin tuna, Isla Wolf, Galapagos Islands- *David Rolla*

The female has done something remarkable; in tropical waters she doesn't risk freezing to death so she has burned up most of her insulating fat and muscle to fuel the production of her young. She will shed up to three quarters of her weight in eggs, reducing her oxygen requirements significantly in the process. Other fish like sturgeons use similar strategies but few do so as spectacularly as this tuna. When the breeding season ends the bluefin will head back to cooler waters where, for a time, although the females will be physically larger than the males many of them are much lighter.

Tuna are not the only warm blooded bony fish. Mackerel are also efficiently insulated with muscles; indeed it this advanced family that the warm blooded mackerel shark group (including the great white and mako) are named after. Their size means that despite their muscle they are top prey rather than predator. The juveniles shoal together but this behaviour is less entrenched in adults, instead they tend to return to deeper water during the day. They have an intriguing ability that alerts them when it is safe to rise. Mackerel eyes are sensitive to shades of blue, particularly ultraviolet, as this is the last visible colour of the spectrum and travels the furthest. It is only when the fish enter water so deep that only ultraviolet light is visible that they stop their journey. As the sun sets the fish follow the vanishing light back to the surface.

When the moon is full enough light enters the sea that several different shades of blue become visible. This triggers the mackerel to descend until again the only colour perceptible is ultraviolet. On a moonless night however the mackerel may venture all the way to the surface feeding on the plankton, unseen by most large predators. But there is one distant relative of the mackerel that can also see in these conditions and the moonless night puts it in a perfect situation to strike.

Swordfish have eyes the size of oranges; the largest eyes of all fish. Like mackerel they utilize muscles to maintain their body temperature but one muscle is particularly enlarged, the one between the eyes and brain allowing both organs to function regularly in frigid waters. So not only can they outpace their prey they can see it early enough to cut off its retreat. Their bill is a quarter of their body length, allowing them to slash and immobilise a mackerel from more than a metre away, reducing the risk to its vulnerable eyes. Occasionally swordfish target more athletic prey, launching from the surface to stab unwary seabirds. They have even been known to attack fishing boats, less successfully, although normally their real target is shoals of small fish hiding under the vessel. One swordfish even became impaled against a submarine, unable to be retracted until the vehicle eventually surfaced.

Swordfish belong to the billfish family; some of the biggest and quickest bony fish. Other billfish, like the spearfish and marlin, are also named for weapons which resemble their bill. The blue marlin is the largest and among the most powerful of the billfish; some individuals can weigh over 900 kilograms and, when caught on a rod, can drag a boat behind them for several hours. Just like tuna the biggest animals are always female and for the same reasons. They also tend to hunt socially, herding prey into a baitball and picking off the stragglers. The blue marlin's bill is substantial, poking a foot or more away from its skull. The bill can smoothly pluck up any prey but the marlin's main weapon is actually its speed. This allows them to bypass the weaker individuals at the perimeter of the baitball to reach juicy healthy prey at the heart of the shoal. The marlins coordinate their attack by raising and dropping their prominent dorsal fins, signalling to each other in a type of sign Even marlin can be overtaken by another member of the billfish family. Sailfish are dwarfed by marlins, even the heaviest weigh only eighty kilos, but what weight they do have is efficiently positioned to create the fastest fish on earth, indeed nothing can swim faster than them. Sailfish travel regularly at more than sixty kilometres an hour, an essential skill for a hunter which has to track nomadic fish throughout the endless ocean. But such a pace is only cruising for this animal.

When a sailfish finds a suitable taget, such as a school of mackerel scad (a type of jack with the same general shape as true mackerel) it begins circling them, testing the miniature fish to show their weaknesses. Any individual unable to keep up with the group soon become obvious. The sailfish accelerates towards its target at a speed that seems almost cruel, a dizzying 110 kilometres an hour. Yet the mackerel scad can still manage to outmanoeuvre the hunter if it can change direction just before the hunter strikes. It is only now that the sailfish demonstrates its reliance on its bill, by tilting its head up and attempting to snatch up the escaping scad l.

Even with its swiftness and extensive agility the sailfish may still take several swoops into the midst of the shoal before it finally catches anything. As with the great white shark such an active lifestyle comes at a cost, a prodigious appetite. But schools of prey significant enough to attract the attention of one sailfish often attract more than one. Harried by the increasing number of predators small fish, including unrelated fish such as true mackerel, draw tightly together in a bait ball, just as they do when hunted by marlin. Such a compact feast of meat, while offering protection to the majority of the mackerel also provides the ideal form for them to be collected in.

The problem is that with rugby player sized fish racing past each other at over a hundred kilometres an hour any collision could be fatal, particularly with that formidable bill. Sailfish cannot lower their dorsal fin in order to communicate as marlins do. Instead, as the adrenalin filled fish prepare to attack they flush their broad dorsal fins with blood. The increase in oxygen causes the fins to change in colour from a glistening blue to purple then through to crimson. The visual change acts like a signpost between sailfish; communicating its forthcoming steering to its companions.

Indo-Pacific Sailfish hunting a school of Mackerel scad, west of Cancun, Mexico- *Rudi Diesel*

When two dozen or so sailfish gather the tension builds as they work together, slicing through the baitball at speed. The colour of the fins provides a further, unexpected benefit to the hunters. The blue and purple tones are similar to the light waves that mackerel use to guide themselves into deeper water. Rather than flee the mackerel are uncontrollably drawn towards the sailfish, mistaken by pure instinct. It is only when the sailfish have finished their hunt and stopped flashing their deceptive fins that the mackerel can finally orientate themselves correctly and escape into the safety of deeper water.

Occasionally conditions combine to bring together not just a dozen or so sailfish but more than eighty. In this situation the billfish are basically touching as they squeeze their way through the bait ball. The increased predators wedge the bait ball tighter and tighter as each individual mackerel tries to avoid its fate. This brings the sailfish within centimetres of each other. Still they continue racing by at break neck speed. Now they are flashing so fast that the colour on their fins is basically flicking on and off. Only in slow motion can the beauty of this marine ballet truly be appreciated.

The scarce resources available in the open ocean mean that both predator and prey are fighting to survive; only failure for the sailfish will take slightly longer to be realised. After an hour any remaining mackerel will have scattered into shoals too small to be hunted down. The sailfish by then will also have left long before in search of their next meal. The only remnant of an hour ago will be scales and fish scraps reflecting through the waves. Over the next couple of weeks the remaining debris will sink down as marine snow to the larvae of the orange roughy, the stoplight loosejaw and the tripodfish; generating the next denizens of the deep.

Chapter 8. Life on the Edge

Crocodile fish at rest, and on the move, Antarctic Peninsula- *Justin Hofman*

Few environments are as unforgiving as Antarctica in winter. The waters are below freezing temperature and the land sheered by winds powerful enough to have carved their mark into the mountains. There are no herds of caribou, no packs of wolves. No reptiles or amphibians can survive in this frozen wasteland. Only the emperor penguin and the Weddell seal remain near the mainland. The cost of existence is severe; enduring on the land means keeping their bodies warm and continually refuelling their energy. Both animals may lose whole populations from starvation or changing weather conditions.

But below the ice life thrives. Krill swarm in their billions, while starfish, crabs and worms scavenge the bodies of dead seals and other debris they can unearth. Part of the secret to their survival is due to the salt water itself. Fresh water freezes at 0°C but salt acts as a mild anti-freeze resulting in much of the Antarctic seas remaining liquid even though it is several degrees below freezing temperature. Invertebrates thrive in these frigid conditions but for fish it is a different matter. No catfish, barracuda or speedy billfish survive here. Until Janurary 2025 it was thought to be off limits to sharks too until a relative of the Greenland shark was surprisingly observed by a baited deepsea camera nearly 500 m below the surface, with an equally unexpected skate also filmed simultaneously moving towards the bait. Most of the local fish are slovenly and belong to a single family of ray finned fish, the icefish. The largest has a head like a crocodile and is one of the key predators for many invertebrates and any other fish tough enough to survive here.

Their extreme adaptations only became apparent to fishermen after they sliced them open. Unlike any other vertebrate the crocodile fish, and other icefish, have clear blood. Originally the transported bodies were considered a hoax as outlandish as any platypus. All other backboned animals, including such odd and distant relatives as the lamprey and lancelet, have bright red blood. The red colour is produced by small oxygen rich cells. Other animals in especially cold climates can have reductions in red blood cells as it decreases the thickness of the blood, making it easier to flow in these pressing environments; but having no red blood cells at all took that to another level entirely. When red blood cells get below a certain level in any other vertebrate, be that a lizard, piranha or human, the brain will begin to be starve of oxygen. If this isn't corrected quickly the animal will perish. The crocodile fish had no red blood cells in its entire body. How on earth was it breathing?

Cold water naturally contains higher levels of dissolved gases like oxygen, but this is not all. The crocodile fish breathes like an amphibian, taking in oxygen through its skin. To do this it has a smooth scaleless body which allows easy absorption. Its slender body also eases this process. A massive heart, four times larger than typical for its body size, propels the precious blood around its body. For even though its blood no longer transports oxygen it still carries vital white blood cells that can destroy any infection threats, and plasma to clot and repair any damage to the body; indeed it has higher blood volume levels than any other fish. These specialised cells still make up only a small proportion of the blood; the rest of it is mainly normal fluid which is at risk of freezing. So the crocodile fish produces chemicals in its body that act similar to man-made anti-freeze, keeping the fluid inside its body super cool forcing it to stay in liquid form. Despite its name this species of icefish is naturally protected from turning into ice.

Being unable to control its oxygen levels renders the fish's swim bladder redundant so it doesn't exist either. Instead it stores high concentrations of oil inside its body, the same method that a shark uses to maintain its position in the water. Getting rid of red blood cells also substantially reduces its iron requirements, potentially the original reason for their loss. It is an extreme solution to an extreme environment. Other fish like the Antarctic toothfish, the largest fish in these waters and a voracious hunter, also have natural antifreeze inside their body, but even it does not have such peculiar blood. Through the crocodile fish's combination of different anatomical features it can survive the most severe weather conditions its environment can throw at it.

The frozen wastes of Antarctica are not the only waters where fish regularly endure such frigid conditions. The northern lands also cope with an annual deep freeze. Unlike its southern equivalent the north pole is not located in a continent surrounded by ocean but instead in a vast ocean penned in by land. This ocean is deep and large enough that fish can dive beneath the increasing layers of ice in winter. As the sun sets, both Arctic cod and herring utilize these techniques to shelter comfortably inside deep fjords. But inland the relatively shallow lakes suffer the full effects of the seasons and the fish here have to be equally skilled in dealing with the ice. Many rely on their eggs surviving the deep freeze stuck to submerged plants while they follow the same rivers and streams that lead them here in the summer to travel south until conditions are fair enough to return again. But in one massive Russian lake the fish use a technique similar to the icefish to survive, and remain active, during the deep freeze.

Lake Baikal is the deepest of all inland lakes, plunging 1,600 metres below sea level. The enormous amount of water here means that any fish that can take advantage of the perilous conditions can really thrive. As a result most of the fish belong to just a few species of golomyanka or oilfish, transparent wrigglers that dart between the icicles after equally unusual relatives of the shrimp. Their transparency is not due to them lacking red blood cells but a different survival technique. Almost 40% of their body is made up of oil, higher levels even than that of sharks or the crocodile fish. The golomyanka is able to rely on this oil, and it alone, from stopping its internal structures from freezing. So effective is this insulation that the female gives birth to live young, keeping more than two and a half thousand fry developing inside her until their bodies are ready for the outside temperatures. The only other significant species of fish in the lake is a relative of the salmon, the Baikal omul, which has to leave the lake to spawn successfully. The measures taken by the golomyanka can prove counter-productive.

Western buffalo bream, Rottnest Island, Western Australia-*Pam Osborn*

In water a degree or so above freezing the animal begins to suffer the full effects of heat stroke. When taken out of the water the reaction is even more visible as the fish literally begins to melt away in front of you. All that is left on the ice is a dried up skeleton with a thin cloak of skin clinging to it, while oil runs from the carcass back to the lake. The price of such specialism can be high.

Specialism typically takes thousands if not millions of years to occur, for an animal to be modified by its environment. Some fish take a more proactive approach and change the environment to suit them. In Rottnest Island, off Western Australia, two species of buffalo bream graze the algae that grow along the coast. The common species tend to congregate in the most productive areas where passing currents result in prolific growth of brown algae. With the change of season however this species begins to starve and has to rely on its fat reserves until spring. The other species, the Western buffalo bream, rely on a slightly different strategy. Their main diet is red algae, which they farm in a similar way to some damselfish species on a reef. The damselfish are known for their prodigious effort maintaining their small territories with their choice tit-bits and aggressively charging any intruders, including human divers. But buffalo bream are much larger and consequently their extensive gardening can be observed by more than just a passing swimmer.

A flight over Rottnest Island in a light aircraft reveals unusual patterns in the surrounding coast that indicate the fish's unusual power. Rows of peculiar pink hexagons run for miles, each with dense green borders. It looks like the result of some unusual geological formation but these distinctive landmarks are entirely due to the western buffalo bream. The flush of pink in the centre of the hexagon is the combined effect of thousands of alga growing together. In other parts of the island the algae clings to existence under the shade of large kelp and other seaweed. But the house worthy bream does not tolerate such inedible plants. It rips them out along with other unwanted weeds. In this way only its favourite fodder is left to flourish. This basically guarantees the western buffalo bream a constant unmoveable food supply unlike the common species which has to travel widely to forage. Such an approach also comes at a cost; time and energy.

The diligent bream cannot afford to share its food with companions and attacks any other fish that venture into its home ground. Consequently a 'No Man's Land' of straggling kelp and other marine algae has sprung up between each territory. Another result of their aggression is the territory's distinctive shape; as the hexagon design allows each fish to patrol its patch and utilise the maximum area of the coast available To retain such productive gardens the western buffalo bream has to recognise different algae and the appropriate action for each; otherwise their carefully tended patches would soon be overrun with inedible weeds; which is certainly not the case. Their competitive lifestyle means that they cannot share their individual behaviours; they cannot develop culture like primates and hence the knowledge learnt by each fish is lost when it dies. Nevertheless they have found a way of creating the marine version of a supermarket, by utilizing the environment to support their own needs.

More than half the world is covered in water and dominated by fish. Few can exert as much control over their environment as western buffalo bream and instead their survival rests solely on the mercy of the seas. While the seas support great shoals of fish they also harbour numerous predators. Other groups of animals like seals and marine iguanas can shelter on land after feeding but an inability to breathe out of water restricts most fish to staying in the danger zone permanently. But competition and specialism has led some fish to overcome even this great hurdle and attempt to escape the water, at least temporarily.

One of the most dramatic examples of this can be seen in the Pacific Ocean. Here shoals of dorado feast on whatever fish they can drive to the surface. Needlefish are especially vulnerable. These are javelin shaped fish that use their bodies to slice their way through the water. Their speed keeps them out of reach of most predators. But dorado are also built for speed. They pursue the needlefish to their limits. Needlefish are typically solitary or in groups of less than half a dozen so they are also outnumbered by the large dorado shoals. The hunters force their prey towards the surface, cutting off any chance of an escape route. The needlefish is trapped by their pursuers below and a thin barrier of air that even the most desperate of fish rarely cross. But with its speed the needlefish can shoot out of the water with such force that they are quickly out of the reach of the dorado. They can travel for a metre or so before their body plummets back into the water, enough to foil most predators. The escape of the needlefish is uncontrolled as soon as it leaves the water. With no ability to steer the fish may encounter and crash directly into another hazard, harpooning a fisherman (or cameraman) in a boat carefully positioned above the shoal. It is a rudimentary escape measure at best.

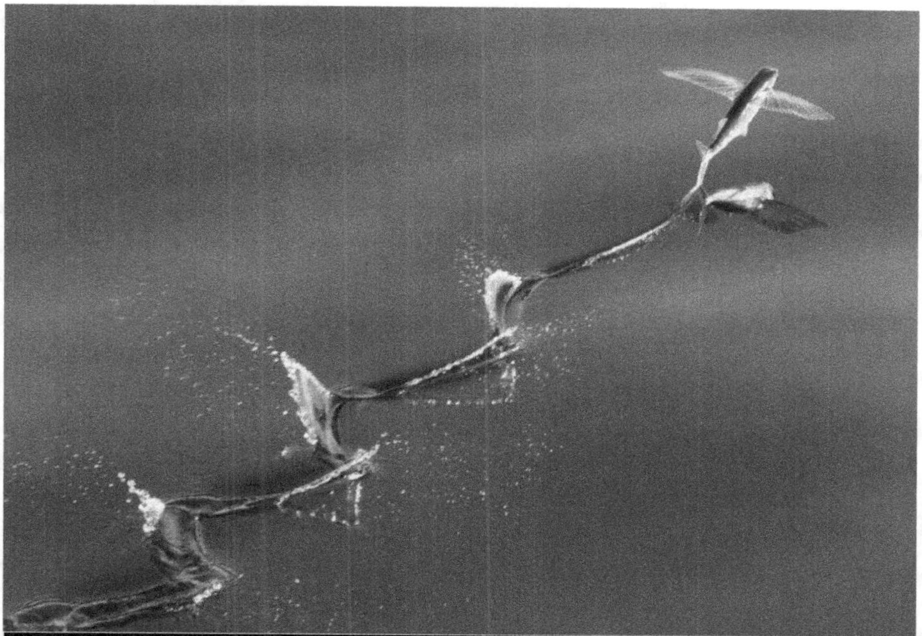

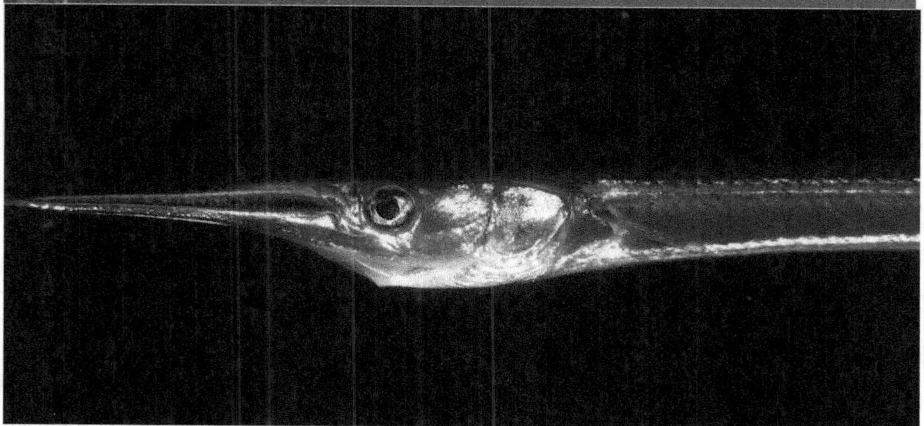

Atlantic Flying Fish, east of Florida, Atlantic Ocean- ***Andrew Dreelin***
Mexican Needlefish- ***Matt Marr***

The sharpfin is a relative of needlefish with a body even further developed to escaping via the air. When swimming it resembles the needlefish except it has a squatter body. Its head angles upwards at its tip so that its nostrils are almost equal in height to the eyes. Its pectoral fins give slight indication of its defensive power. They are oversized and appear more of a hindrance than anything else. The sharpfin may not be the slowest fish in the ocean but under the water it is far from athletic. That is because swimming is not their pectoral fins main purpose. Like needlefish the sharpfin are rather attractive prey for predators. Their large schools make them an even more tempting target for passing dorado.

As the dorado draw closer something innate seems to click inside the sharpfin and it springs into life. The shoal rises so that they are immediately below the surface, each individual flicking its tail at full speed. Spacing is critical. Most prey fish gather tightly together, using sheer force of numbers to overwhelm their predators. Were the sharpfin to do this their defence strategy would likely result in collision if not more serious injury. Instead they separate as much as they can in a few seconds. Beneath the very surface of the water is the highest concentration of oxygen which allows the sharpfin to increase its energy level and speed before it does the inevitable. It slices through the water and releases its fins! Only now that they are above water do the cumbersome fins prove their merit. By applying pressure the fins are held in place and the fish begins to glide. Watching a sharpfin its floating ascent looks effortless. In reality the fish is slowly suffocating. Once its gills leave the water they rapidly dry and temporarily stop functioning. After a few metres the sharpfin soon outpaces most pursuers and drops back into the sea. But dorado have speed that other hunters lack, pushing the sharpfin often beyond its limits, sometimes even catching unlucky individuals as they smash back into the water. The sharpfin is among the most primitive of a family of highly efficient gliders. Not without reason did their discoverers call them flying fish, even if technically they don't actually fly. Even so some species may travel for more than a hundred metres once airborne. But to travel that far requires significant advancement to the basic design of the sharpfin flying fish.

In two wing flying fish the pectoral fins are even longer than those of the sharpfin and have the ability to lock in position while gliding. These are some of the most common species of their kind, gathering in vast shoals that lift together almost simultaneously but they are not the most advanced. That honour goes to an Atlantic species, the four winged fish. Their pelvic fins are also adapted for the air, although not as extensively as their pectoral fins, providing greater agility. They still start their escape just below the surface to raise their oxygen levels before leaving the water. But not all of their body will leave it. The tail has an overhanging lobe that hits the surface at an impressive rate of sixty times a second, allowing the fish to skate over the water before commencing gliding.

The four winged species stays in the air longer than any other flying fish resulting in it being submitted to increased physiological stresses. Once in the air it begins to lower its oxygen consumption by simultaneously dropping its pulse and blood pressure. Its eyes have to temporarily refocus so that it can avoid crashing into another before adjusting just as quickly as it re-enters the water. This most agile of gliders may have travelled for two hundred metres before plunging back into the sea, providing at least temporary escape from ravenous dorado.

Inland there is a group of little known fish that *can* actually fly. In the Amazon predators are just as common as in the open ocean and patrolling storks could quickly snatch any casual gliders. But hatchet fish do more than glide; their short fins are capable of flapping and producing sustained flight, giving them some chance of steering away from predatory birds. This skill has another purpose. In the heat of the equator small ponds can quickly evaporate, driving shoals of fish uncomfortably close together. The increased stress drives typically passive scavengers like piranhas into a feeding frenzy; in the excitement even turning on each other. While the pools churn the hatchet fish can simply fly and seek out safer waters.

Common Hatchet fish- *Lifeonwhite*

Although they are gutsy fliers hatchetfish can simply not compare with birds as masters of the air. A heron can easily flick a leaping fish into its mouth while a hawk can change direction mid-flight and spear a hatchet through the middle. Nor, for that matter is their brief ascent even close to as controlled as the glide of the flying fish. Escape in the air for the most part relies on little more than luck. Hatchetfish have another disadvantage; light is deflected as it enters the water and their eyes are naturally adjusted to the distortion. As soon as they leave the water however they are momentarily blinded and experience the same visual difficulty as someone putting on the wrong prescription glasses.

There are opportunities for food as well as escape above water though, if you have the means to access and see it. Bordering much of the coast of Indonesia, Asia and northern Australia are vast tracts of mangroves. These formidable trees are able to survive daily submersion in waters so salty that it kills any of their competitors. Once the tide recedes the mangroves are submitted to a roasting by the equatorial sun and have to rely on every ounce of water saved up in their roots. The tree's roots jut out above the high tide mark allowing it to access fresh water when it rain. The labyrinth of roots is equally important below the water where they snare any passing debris and consequently attract large numbers of scavenging invertebrates. Low tide allows land crabs and insects to join the feast but the current often returns before they're ready. Large shoals of jacks, tarpon and even occasionally barracuda hunt below the surface, snatching up any drowning or straggling prey. Even more rarely a saratoga may venture into the estuaries, leaping to pull hermit crabs from lower branches. But only one fish has the ability to hunt above the water without ever needing to leave it.

King tides often result in fish carcasses being abandoned on the mangroves branches, attracting vast swarms of flies. Archerfish begin to gather, watching the insects intently. Sometimes the flies are low enough that they can be easily snapped from the surface but often they are tantalisingly out of reach. The archerfish responds to this in an extraordinary way. During the middle of inhaling it quickly jerks its mouth back and the resulting pressure acts like the piston in a gun. It shoots a stream of water out of its jaws towards the fly. But unfortunately its meal is not where it appears to the archerfish. Light refracts any image above the surface, distorting the fly's actual location.

◄Spotted Archerfish, Ross River, near Townsville Queensland, Australia, *Alec Karcz*
◄Great Blue-spotted Mudskipper, Kyushu, Japan- *Suzu Sato*

Remarkably archerfish can learn to compensate for this by aiming its attack slightly to one side of the fly. Consequently the spray often makes a direct hit, knocking it straight into the Archerfish' open mouth.

A juvenile archerfish may take several attempts to hit the target. Each time it sprays it inadvertently sends a disturbance vibrating around it, which is soon picked up by the delicate lateral lines of other archerfish. Half a dozen or so will soon gather, all competing for the same single fly. Each time they spray and miss the additional disturbance notifies other competitors. There is no guarante that the fish that hits the fly will get to eat it. Some archerfish get a bit more aggressive as the group grows and begin to leap for it. While the fish are aware of refraction when spitting at prey, leaping at the food seems to throw them. Often they will jump right past the fly, pulling at leaves as they pass. Occasionally the archers veer so far off course that they crash into a branch.

Even archerfish have to abandon their feeding grounds when the tide retreats. But now is the time when other fish venture into the swamp to feed. Small bubbles begin to froth out of the draining sand, followed by a pair of eyes which rise up like periscopes. It is a goby that spends most of its life on land, a mudskipper, and only with the tide gone will it feed. It is an unusually poor swimmer and could be outpaced by any of the larger fish. Even so like all fish if its skin and gills were to dry it would die. So the mudfish is restricted to sand damp enough to sustain it. At times it will even roll over, uniformly distributing the moisture around its body. The protruding eyes are most at risk of drying out so conveniently these can be popped in and then back out of the head, saturated once more. At midday even this may not be enough and the mudskipper is forced to travel back to the water to keep moist. But it is highly selective where it goes, only heading for isolated pools too small for significant predators.

With minimal competition mudskippers have been able to adapt their body shape to harvest basically all food sources on the estuary. In Indonesia, where the family's diversity is at its greatest, several species can share the same area. The larger kind scoffs up lounging crustaceans and molluscs while lesser species are generally more opportunistic. Closer to the estuary herds of bulky mudskippers plough up the mud after marine algae. This can lead to fierce competition as males are forced together; when they posture vigorously with their fins and snap at each other. But with the constant threat of immersion such fights are short lived. In order to feed more efficiently some mudskippers construct a circular dam around them then gulp mouthfuls of sand behind it, searching for plankton or any other nutrients they can extract from it.

Once the water returns the mudskipper has two options; some dig deep burrows while others start to climb the mangroves, sheltering among the humid leaves. It is here that the mudskipper's transition to terrestrial life is revealed as being comparatively recent. It has no claws for those individuals that dig, and no toes or limbs to support its body as it ventures into perilous heights. Instead, like most gobies, it has fused fins that allow it to cling partially to the tree. This is a relatively slow process so the mudskipper has another, more urgent, means of moving. It can push back into its fins and jump, hopping uncontrollably so unexpectedly that it may scare off any potential predators, say a small bird, until it plops back into the sea. It can even continue its eccentric skipping motion on top of the water surface so that it is effectively walking on water. This hopping could cause a mudskipper to be mistaken at first for a frog until you realise that any amphibian would soon perish in the salt water, one thing the mudskipper has no risk of. But even so it is safest out of water and once landed in it quickly resumes the long process of climbing back out.

The mudskipper is not the only fish to skip over the land. There is a certain catfish in India and much of Southeast Asia that frequents pools which regularly evaporate. Just as in the Amazon the drying pool triggers fish to panic and turn on each other. The catfish has some protection for this, the stiff spines imbedded in the front of its chest fins. The spines are long enough to stretch the length of the vulnerable fin. But they provide more than just defence; they are sturdy enough to allow the catfish to drag its body through the clinging mud. Eventually the pool may shrink enough that the fish has to abandon it altogether. Unlike the mudskipper its fins are barely affective above water; the entire weight of the fish is supported by the two spines. Walking on two legs is difficult enough; every time a limb is moved a human is at risk of losing balance and toppling over. Relying on a pair of inflexible spines is even more challenging. There is no flexing joint in the rigid spine equivalent to a knee so each time one is moved it sends a sudden jolt through the body. Having the rest of the body dragging on the ground behind it further adds to the difficulty.

The movement could hardly be called smooth even when compared to the awkward hopping motion of the mudskipper. Nevertheless this unusual jungle stroller is called popularly the walking catfish. Overall spines are not the most effective form of mobility. It is only the fish's modest size that makes the feat possible at all.

Some fish are efficient walkers, but back under the water, and not due to defensive spines. Sea robins, such as the Pipperfish, are distinguished by their fins that distend like a pair of oversized hands. They can swim but they prefer to spend most of their time scuttling along the ground. The fin not only resembles a hand it can be used in a similar way. The front three rays on its pectoral fins are loose and can be moved independently. The sea robin uses these rays like fingers, feeling through the sand and around rocks for any potential food. These special fins are also covered in taste buds, allowing them to act like a second tongue; tasting any potential prey without it needing to enter its vulnerable mouth. When satisfied with the taste the sea robin uses those same fin rays to shovel the food inside its jaws.

In the waters off Tasmania and Southern Australia are found even more unlikely pedestrians, handfish. These specialised anglerfish have developed both pectoral and pelvic fins into flexible muscular limbs. These fins can still be used to swim but their primary purpose is walking. The hand-like fin lacks the taste buds and separate rays that the sea robin has, but it is sturdy enough to hold the body high above the substrate.

Like all anglerfish it isn't a particularly great handicap for a handfish to be slow; they don't have to catch up to their prey- by waving the esca around on their head prey comes straight to them. The five identified species of handfish plod across the seabed like tortoises, except for one crucial difference. It is one thing to be able to carry your weight on fins that consist of only rays and no bones when you're supported by the high buoyancy of water around you. On land there is no such support and even the handfish would soon collapse under its own weight in these conditions. The first fish to enter land would have done so only briefly.

Pipperfish, Cabo Cope-Puntas del Calnegre National Park, Spain- ***Alberto Carrera***
Red Handfish, Derwent River, Hobart- Dr John Turnbull

While there is food on land for the likes of mudskippers and archerfish it has another advantage that may have originally caught the attention of the first fish to leave the water, well before either of those groups appeared. In the sea vast hordes of animals gorge themselves on any eggs floating in front of them. Even if the eggs lie buried in the sand a hungry snail or ray could easily smell and snuffle up the edible morsels. Nor are predators the only danger. A change in current could deposit juveniles many miles from food and acceptable temperature. On the shore though the developing egg faces comparatively less predators and, if deposited safely, the fish inside can grow to a healthy size until it is fully ready to face the mercy of the ocean's currents. But the coast is as much a hazard to the adults as it is a safe haven for their young.

The coasts of Newfoundland rise gently from the sea; the beaches lined with round pebbles that stretch for several hundred metres. The cobbled stones are a result of the relentless action of the waves; grinding and rolling the pebbles together until they are as smooth as marbles. The bare beaches appear lifeless. But clinging on in a few protected areas are clumps of kelp and algae. The plants are fiercely battered by the waves, often ripped out by them in fierce winter storms. In high tide the ravaged coastline becomes a little more accessible but there is always a risk of stranding as the water retreats. Those small fish that can keep their own under such conditions are basically predator free. Capelins in particular, a relative of the salmon, spend most of their life further out at sea; but when June arrives and they begin to spawn they start to head towards the coast. Capelins that live in the Grand Bank, as well as their relatives in Europe, live in waters temperate enough for them to lay their eggs on the seabed. But the majority of Pacific capelins live in waters too cool for their eggs to survive and consequently take a rather more extravagant protective measure.

The capelins take advantage of the high tide and glide on the crest of the wave towards the land. As the wave gushes back into the sea the fish are left drying on the shore. But the stranding is deliberate.

Thousands of capelin surge on top of one another. A refreshed group seems to come up with each wave. The female begins to squirm, releasing up to sixty thousand eggs which begin to seep beneath the pebbles. The male quickly squirts his sperm over the eggs and catches the next wave out. The female remains a little longer until she has laid all her eggs. After such an effort she may die. Gulls and sea eagles fill their beaks with both fish and eggs. But most of the eggs will be washed with the incoming water ever deeper into the beach. Here they will be safe until they hatch a few weeks later.

Off the coast of California are found one of the only other fish that risk spawning on land. Grunions are spiny ray finned fish which, if possible, take even more care with their young. For most of the year they steer close to the coast but in spring they begin to follow the very limits of the tides into land. Like the capelin they lay their eggs out of water but they are much pickier in their timing. The entire species spawn on the same couple of nights, which increases the survival of their progeny exponentially. They rely on the only cue visible across the vast Pacific Ocean; the moon. A full moon brings the tides to their highest extreme and the Californian grunions follow the torrential waves on their way to the sand. But they do not spawn now. Any eggs laid now would dry out so far from the ocean. Two nights later conditions are perfect. A solitary grunion, typically a male, washes onto the sand and by compressing his swim bladder creates an unusual squeaking noise. His voice is answered by a chorus from several shoals converging towards him, onto the moist sand.

The females have the most arduous task in bringing forth life. They travel forward on a wave then squirm into the sand as the water retreats. By wriggling their tails they half bury themselves in the damp substrate and get into position. The males come tumbling after with the next wave; encircling a female and releasing a stream of sperm before being washed back into the sea. The female remains, producing eggs for other competing males with such determination that she may perish of exhaustion soon after or even die during the process. By the end of the night there is no indication of what has occurred; the millions of eggs shed during the evening are safely sunken beyond reach of any marine predators. Some land-based hunters like raccoons and black bears can sniff out such a bounty, even so they make little dent in their numbers.

Such a ritual continues until six nights after the full moon. Then all is quiet for nearly a month and the adult fish disappear from view once more. But the apparent emptiness on the beach is deceptive. Inside each successfully fertilized egg an embryo is making preparations to hatch. Not all of the eggs remain on land, some were washed out with their parents and a few of those may survive long enough to develop into an embryo. On the beach their siblings are ready to enter the sea but they have a conundrum.

Pacific Capelin washed up on shore, Baranof Island, Alaska, USA- ***Tim Sorrill***
Spawning Californian Grunion, San Diego County, California- ***Tobias Hays***

Once they break out of their egg they lose the protective moist shelter and, if they've got their timing wrong, will dry out almost immediately.

The full tide that comes with the full moon is their best chance of survival but the unhatched fish have no way to see either tide or moon. But they can feel. The crash of monstrous waves against them begins to stir the babies. The sand itself, which has held them safe for so long, begins to crumble at this marine power. The eggs are swept up and cast around as if they were in a washing machine and this fierce movement triggers the young grunion to hatch and take on an adult existence. The grunion eggs drifting freely in the ocean will never be subjected to this experience and consequently will fail to hatch, dying inside their egg. They have suffered the penalty for their parents' audacity at leaving the sea to breed.

Recently a third species of fish has been recorded in Japan as engaging in the extreme behaviour of leaving the water to breed, the grass puffer, which has volatile poison to protect itself while out of the water. The fact that even such highly toxic fish engage in this risky behaviour emphasises how hazardous the sea is; and the risks that some species will take when they need to. Other fish venture into perilous situations not to breed but to feed.

In Africa's rift valley hidden subterranean activity releases small streams of volcanic material to the surface. Like the deep sea vents in the oceans these waters are boiling but they are also high in essential minerals which support vast numbers of a few keen specialists. Where the water cools a little algae takes hold and grows quickly due to the ample amounts of both sunlight and nutrients. As the alga grows it begins to block the flow of the boiling water, cooling the water beyond it enough for it to be colonised by an enterprising fish. Cichlids are common in the valley and one, the tilapia, grazes algae wherever it can be found. In the waters seeping from the vents they quickly clear out the most accessible algae. To survive the tilapia must venture briefly deeper into the springs to graze the untapped growth. The thick level of algae significantly cools the water but even so it is still hotter than most animals can stand, at more than 50°C. For a cold blooded fish the tilapia takes an awesome risk, one that many will not survive. As they take a gulp of algae they also swallow gases that begin to damage its respiratory tract. The fish are being suffocated but they are also being poisoned. Most tilapia will prudently leave this part of the springs after taking just a single mouthful of food. Others in their hunger may stay for another greedy mouthful and this may prove their undoing. Helplessly they rise to the surface of the lake as their swim bladders expand with toxic gas. The resulting carnage will decay, boosting the alga growth and tempting yet more fish into these dangerous waters.

The noxious waters where tilapias live are highly alkaline, high in nutrients and waste chemicals like urea. Worldwide most fish prefer water that is either neutral, with a ph of 7, or slightly alkaline. But much of the water in the Amazon River basin is mildly acidic. During the early 1900s cardinal and neon tetras were exported from here across the northern hemisphere in their thousands. They were priced to match their beauty and this encouraged a few entrepreneurs to attempt breeding the species. Despite the skill and expertise of those charged with their care no one in France, England, or many other parts of Europe could encourage the tetras to mate. Nor could ichthyologists in North America. In Saxony, Germany, however breeders were effectively turning out generation upon generation of the gaudy fish. Tensions during the early part of the century were already running high between Germany and other nations. When the waters were tested it was found that the tetras were breeding in waters with a ph of 5. The water in Saxony was close enough in ph to that of their original South American home to enable the tetras to breed.

Acid is as much a barrier to life as alkalinity. In addition to affecting respiration, at high concentrations it burns into skin even that of people. The waters of Saxony and the Amazon are not severe enough to produce this reaction. Surrounding forests and plains produce enough debris to partially neutralise the water. In Western Australia there are a few special places where the water is not neutralised by either of these processes. Instead, coastal woodland entraps small ponds, sheltering them from the surrounding desert. Sporadic rainfall and fog keep the ponds continually low. Occasionally the pools may evaporate completely. Constant drought also stops neutralising leaves and waste being washed into the pools. Consequently acid levels build up excessively. The pools can have a ph down to 3, similar levels to that found in some diet soft drinks, concoctions capable of beginning to corrode teeth in a matter of days. Most fish would die in these conditions. Yet the pools hold one fish which spends its entire life here and indeed survives nowhere else.

Squirming in the drying mud is a creature that looks like a scaly salamander. But Australia has no such amphibians and scientists identified it as a scaled galaxias, part of the same group of fish that undertake annual migrations to New Zealand rivers. But galaxias have no scales and the salamander fish was given its own family to recognise both this and a number of other important features. Part of the reason why it resembles an amphibian comes down to a curious modification behind its skull. In the acidic water any loose pieces of skin would soon clog with toxic substances and begin to burn. This includes the loose stretches of skin surrounding the eyes, which allows all other vertebrates (with the exception of owls) the flexibility to move their eyes. So, the salamander fish has lost both this essential patch of skin and the surrounding muscles, leaving it with a fixated view. As a compromise this fish has a form of neck so it can bend and tilt its head around. But all is not what it appears. This flexible fish has none of the structure or the necessary musculature to make up a true neck. Instead, the salamander fish has taken the risky move of extending the gaps between several of the vertebra behind its head. In any larger fish such a gap would snap its spine but at only six centimetres long this is not a problem for the salamanderfish. The value of this false neck doesn't really become apparent until spring.

Under the increasing heat even these small pools don't last long. Using their muscular fins to pull themselves across the bottom the little salamanderfish begins to snatch up the last available food. They start tracking the remaining moisture into the peat. They wriggle their flexible spines until they come into contact with soil deep and damp enough to support them even as the ponds dry above them. A secretion of mucus helps to further limit fluid loss. It will be a long wait for the curious creature. Throughout the summer and most of the autumn it will remain unseen underground; breathing through their skin in a state of dormancy. Winter, and the accompanying rain, brings a much needed relief to these creatures.

As their burrows fill the fish tilt their heads, spying the fairy shrimp that begin to hatch from eggs lying in the dust. This ancient type of crustacean is one of the few invertebrates that can survive these conditions and also make a convenient meal for the famished fish. The salamanderfish begin to make up for their time resting, with gusto. A pair shuffles close, pressing their sides together. The male's vent is modified to allow it to introduce sperm into the female. In the peat filled water the eggs have an increased chance of being damaged by the acidic conditions but internal fertilization removes that risk. When the female delivers her eggs, about a hundred and twenty of them, they will have developed enough to stand a fair chance at survival under such hazardous conditions.

Part of the difficulty for salamander fish derives from it living near the coast; where the pools can soon be inundated by saline water. In Central America a completely unrelated fish also has to deal with the ravages of living in brackish water. Four eyed fish lives in swamps that can heat up above 30°C, enough to cook most other fish to death. But the four eyed fish is especially hardy and that extends to its diet as well.

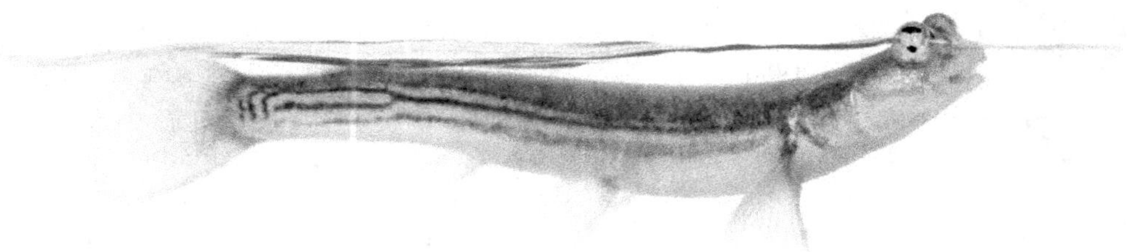

Four eyed fish- *Lifeonwhite*

It is a supreme opportunist willing to consume anything, whether the potential meal is washed up by the tide or fluttering above it. It feeds as much above the tide line as it does below it and that provides it with a visual problem; that of adjusting visually between both a dry and moist environment. The four eyed fish has a more than adequate solution; changing the shape of its eyes so it can simultaneously focus above and below the surface. It rests like a crocodile with its eyes half submerged. The eyes stick out and are divided horizontally with each section possessing a separate cornea and retina, allowing them to focus clearly independent of what the other half of the eye is doing. This effectively provides the fish with split vision; allowing it permanently clear vision regardless of where it is looking.

While the adults are well adapted to these swamps any eggs they laid here would likely be gobbled up or destroyed by the constant tide. They have found a way to protect their young that is even more effective than that used by the grunion. Four eyed fish belong to a group called killifish, many of which give birth to live young. To do this they need a mechanism of fertilizing the egg internally and the male four eyed fish has just this. Their anal fin has been modified into a specialised organ capable of releasing sperm into a female, a gonophodium. One of the faults of this fin is that it permanently points in one direction, either left or right. This limits the male to only being capable of breeding with females which can fit him.

As a result of their unusual breeding strategy the four eyed fish has divided into two populations, depending on who they can physically copulate with. In large schools this is barely a problem but if only a few individuals survive an unusually harsh year they may not be able to repopulate the area. Those females that are fertilized keep their eggs inside specialised internal bags, follicles. Nutrient passes through the female's body into the follicles and from there directly into the eggs. Any waste the embryo produces leave the eggs via the same route. In this way the follicle act similar to a mammal's placenta with the chief exception that the four eyed fish remain inside their egg throughout their entire development. It is only as their mother gives birth that they finally break out of their egg.

Live bearing has developed in several families of killifish independently. In caves in Yucatan, Central America, a species of molly survives conditions as pressing as that of the salamander fish. In the darkness sailfin mollies have to rely on their advanced sense of touch to detect the sparse prey. Absence of sunlight results in the cave mollies glowing an unearthly white, due to a significant lack of Vitamin D. Added to this, dripping into the molly's isolated lakes each day are drops not of water but sulphuric acid. The adult can survive such severity but an egg would quickly dissolve here. So the cave molly's eggs develop and hatch inside her. Unlike the four eyed fish the gonophodium of the cave molly, indeed in the majority of live bearing killifish, is flexible so the molly is not as restricted in its choice of breeding partner. In conditions where there is only enough food to support a few molly this basically ensures the species survival. But this advanced form of child birth does not necessarily imply that the molly has any great maternal instincts. Feeding your young inside your body until they are ready to hatch saps a great deal of energy from the mother. As she gives birth she seeks to recover some of that energy. Her young rises from her vent individually, launching straight to the surface to fill their swim bladder with air before they begin to sink. Once they have taken their first breath they can control their own buoyancy and outpace their hungry parents. But some may not get that opportunity. As soon as her young is released the mother turns on them trying to snatch up as many as she can. Those that do survive will not stray near their mother until they have grown considerably.

Sailfin mollies also live outside caves, in wetlands scattered across Central America. Here they are bright, vigorous fish but they also face a related competitor, the Amazon molly. This species is named not after the great rainforest (which is some three thousand kilometres south) but instead after the mythical tribe of women in Ancient Greece. This tribe was renowned both for its warfare and being composed entirely of women. The Amazon molly may be a peaceful vegetarian but just like the legend they are almost entirely female. This species is thought to have arisen from a one off mating between two of the neighbouring kinds of mollies that, unusually for hybrids, was still able to reproduce. Only one or two Amazon males have ever been found and they were sterile. So, the female is forced to seek a mate from another species of molly, like the sailfin. She requires sperm to start her eggs developing but after this she ejects the sperm, meaning any of the resulting offspring will be a clone. The male she has tricked into mating with her has basically been cuckolded, without another male even being required.

In the sunlit waters the sailfin molly is free to produce one of the most spectacular and competitive courtship routines of its family. As soon as a male begins displaying in front of a female a competitor soon appears and they begin to fight. Although they frequently court and mate with any of the other species of molly the male sailfins usually ignore the Amazon females. That is until a female sailfin turns up. The more experienced, and wily, male begins to court not the female of his own species but the Amazon. His rival does not ignore this change and he too begins chasing the Amazon female. As he focuses his courtship on this previously ignored relative the older male quietly makes off with the preferred female. In this way somehow the Amazon molly has survived not just as a one off individual but as an entirely separate species.

The live bearing properties of so many killifish species allows different variations to appear in a short amount of time. In aquariums this has made it easier for fish breeders to deliberately select parents of species, such as guppies, with desired characteristics and develop these over generations. In the wild this selection is just as active. Small pools become isolated from the main rivers occasionally and in these conditions most potential guppy predators quickly starve and die. Under these conditions the remaining guppies are under less pressure to blend in with their environment and the more brightly coloured individuals have a higher chance of attracting a mate and passing on their genes. After a few generations this trait can spread through the entire community. If a flood comes into the area the pool will resume its connection to the river and predators will return. In these situations the guppies that are more camouflaged will again begin to have a higher chance of survival.

Some killifish in West Africa endure rather more complex conditions. Even in the jungles rain is not guaranteed and only the largest lakes and rivers keep most of their liquid. The killifish here tend to prefer isolated less competitive waters and these often disappear altogether and the fish die with them. But their eggs are hardier and can remain for several years in the dust awaiting any rain. The crater formed by an elephant's footprint may contain several of these durable eggs and when the monsoon finally arrives it may be the only free standing water that the killifish ever see. To avoid mating with their siblings they may have to skip across the jungle floor until they find a stream with unrelated inhabitants. Not all of them make it yet enough breed every couple of years that several species of killifish across Africa have adapted this strategy.

The most adverse environments inhabited by any killifish are the mangroves of Central America. Here constant drought can lead to toxic build ups of sodium and other chemicals while in the wet season the mass inundation of fresh water into the estuary forces the fish to reverse its process of osmosis. The continual spray of salt water from the surf into the edge of the mangroves further affects this fish's internal processes. As you travel deeper into the mangroves things become even more interesting. When the dry season arrives here several pools are cut off from the sea entirely, along with any killifish in them.

Mangrove killifish are some of the most adaptable survivors. These extreme fish were the first, and so far the only fish, to journey and survive into outer space, capable of dealing with conditions like zero gravity inside their small tanks and still orientating themselves and feeding. There are few fish, or indeed any other type of backboned animal, that can survive the range of conditions that mangrove killifish

do. But even they cannot remain in the turbid water stagnating in the swamps. They begin to crawl out of the toxic water.

Mangrove Killifish, Guadeloupe, Caribbean- ***Jean-Paul Cicéron***

It may be weeks until they can return to their original territory. Any more favourable areas of swamp are already filled with killifish and they will not permit invaders without a fight. So those stranded killifish take the desperate decision to leave the water altogether. They begin climbing into the trees where the surrounding foliage may be moist enough to keep them alive for a few weeks. But some killifish find more effective accommodation.

Many mangroves shed their leaves to get rid of excess nutrients picked up from the sea. While some of these fall in the water, others get caught between the branches. Here they begin to rot and soak up any rain. But these decaying heaps do not last long. Lines of termites work their way through the partially submerged trees, picking up any rotting vegetation and carting them back to their colony excavated in one of the mangroves. The termites don't eat the leaves directly but instead feed on the fungus that grows on it. The fungus relies on a stable and humid environment in order to grow. To keep the temperature stable the colony is intricately constructed and contains several bore holes to provide adequate air flow. At night the cool spray from the ocean blowing through the trees is enough to maintain the colony's humidity level. These are ideal conditions not just for fungus and termites but for a fish seeking refuge from the elements. Yet such a fortress is formidably armed by hordes of soldier termites. Its mucus laden skin acts as some protection as the killifish pushes past the initial defences. Once inside the colony the fish is safe. The soldier termites for the most part do not venture far into the nest and the worker ants' jaws are too small to harm it. In order to take full advantage of these conditions the structure of its gills and skins change, allowing it to siphon out and process the oxygen from the surrounding air.

The humid conditions are enough to keep the mangrove killifish alive. But it cannot eat either fungus or termites; indeed it is unable to feed at all out of water. The fish is living purely on its fat reserves. It may remain on its own out of the swamps for ten weeks. Crucially it is unlikely to encounter another mangrove killifish until the return of the wet season and were it to wait till then to breed its young would not have enough time to hatch and mature until the drought returned. But the mangrove killifish, hardy as it is, is incapable of leaving its shelter for long enough to find a mate. To have a chance of raising healthy offspring it has to bring them forth on its own. Inside the fish a remarkable fusion of the reproductive tract has taken place. The fish is simultaneously male and female, like the hamlet fish. But hamlet fish are incapable of fertilizing themselves. Mangrove killifish on the other hand have a specialised organ, the ovotestis, where eggs and sperm run into the same sack and are fertilized. The only other fishes known with a similar ability are the deep sea tripodfish. Both have independently stumbled upon a remarkable solution to breeding in some of the most extreme and isolated situations.

The muddy banks of many of South America's rivers are also prone to stagnation and are filled with fish rising to take a breath of fresh air. As well as catfish and electric eels more ancient fish are also driven up. Swamp eels. They expand their mouths like bellows as they rise to the surface. The entire lining of the throat has been modified to absorb oxygen. In the process an array of scents enter their nostrils from the surrounding river-lands and forest. This mostly finless fish resembles an eel but belongs to its own separate family, most of which are as adapted to a terrestrial life as they are in the water. Although it can reach a metre the swamp eel easily wriggles through the mud, scoffing up beetles and other small invertebrates in one swift movement. Frogs and small rodents are also common but are rather more difficult to deal with. The fish seizes a leg and begins to spin madly like a crocodile in a death roll until the limb is ripped right off. With this technique it is able to dismember its meal and finally enjoy it.

Swamp eels are nearly as at home out of the water as mangrove killifish but they are unable to cope with such a range of conditions. If pushed they can remain on land for as long as six months, provided they do not dry out completely. The fish also have peculiar sex lives; nearly all are born female, only changing into males at four years of age. Part of the key to their success is their gills which unite below the throat to create a kind of mono-gill and increase their capacity in the process. But their offspring do not find respiration so easy; they are incapable of breathing surface air and their long bodies impede the process further. When they hatch their bodies are wrapped in pectoral fins so massive that they appear like flippers. As they unfurl the youngsters begin to wave these fins frantically, driving oxygen rich water into their gills in the process. After a fortnight the swamp eel's increasing throat size enables it to start breathing air to supplement its oxygen supplies. The pectoral fins that were once vital for the infant are now redundant. Overnight the fish sheds them and is able to wriggle overland as easily as any normal eel.

There are fish in these rainforests that are even more ancient than the swamp eels. Their overall lithe body shape is similar but it is their fins that set them apart from any other fish on the continent. Rather than consisting of a network of underlying rays they are fleshy and rely on a supporting bone structure similar to our own arm to move. This isn't immediately obvious with this species as the fins have been modified into whip like struts. This creature has as much trouble living here as its competitors and has its own way of surviving in the turbid waters. Instead of the swim bladder of many other fish this species has a pair of large sacks that it uses to inhale air. Indeed the South American lungfish relies solely on its lungs to survive; its gills are basically redundant and if prevented from reaching the surface the fish could easily drown.

Juveniles and developing embryos in their eggs are also only able to breathe surface oxygen. This causes complications because the South American lungfish, like most other fish, lays its eggs under water. The male remains with them and as he guards his offspring his body begins to change. Pink outgrowths begin to ripple from the base of its pelvic fins. These odd growths resemble gills but they are not for the benefit of the parent. Rather than filter oxygen out of the water these gills actually release the precious gas out of the father's blood stream in amounts efficient to aerate the eggs and young. The outgrowths are effectively an inbuilt bubbler system. By this continual transaction the father is able to keep his young alive until they are capable of breathing at the surface.

The South American lungfish is as unfortunate in its choice of habitat as mangrove killifish and often the swamps where it live can dry up completely. Many other fish either migrate or perish- relying on eggs buried in the mud to hatch during the next rain storm, but the lungfish is not so restricted. Breathing is not a problem in this situation but drying out is so the fish buries itself in a cocoon where it can last, at most, for a few months. Some of its relatives, across the Atlantic, can survive far longer than that out of water.

There are four kinds of lungfish scattered across Africa. The spotted lungfish wriggles through the swamps of the Zaire basin. Its fins are still delicate but conspicuously stouter than those of the American species. It can swim, swaying like an eel when it does so, but it more typically drags itself with its fins regardless of whether it's in or out of water. Females and juveniles travel effortlessly throughout the basin but males are more territorial. They dig and guard the same burrow for several years, observing the developing young sheltering inside the make-shift home.

The spotted lungfish also keep watch for passing prey, snapping at any passing frogs or small fish. Many of the creatures they catch are too large to be swallowed whole so the lungfish has to continually chew its food, spit it out and chew it again until it has been masticated enough to be swallowed. Areas rich in prey result in more valuable burrows which are likely to be fought over viciously.

This territorial behaviour becomes problematic when the rains fail and the swamp begins to shrink. Females take no part in constructing burrows and have no hesitation leaving altogether for the deeper lakes and rivers that remain year round. Likewise the juveniles and any males lacking a burrow flee to the permanent waters. They will not return until the rains return. The territorial males remain as the water drains away until eventually the only water remaining is deep inside their burrow.

South American Lungfish, Cayenne, French Guiana- ***Vincent Rufray***
West African Lungfish, Tanguiéta, Benin, Africa- ***Sadam Toro***

The lungfish can come to the surface to breathe but he cannot eat on land. He will have to wait the few months or so until the swamps are submerged again and he can resume battling for his burrow and snapping up prey. It is a tough challenge for the males but it is a rare year when the spotted lungfish's burrow dries up completely.

The West African lungfish is not so lucky. Almost annually the wetlands where they live do dry up completely. It is not only the territorial males that are caught in this situation, so are the females and the young. To outlast the dry season the fish begins to dig, pointing its head down and writhing into soil deep enough to still hold some moisture. While digging the fins are held close to the body to avoid injuring them. In places the substrate is so hard that the fish has to chew it and excrete it through its gills. As the mucus from the lungfish presses against the burrow the dirt turns to mud and begins to thicken, then harden, against the fish's body. But the lungfish continues burrowing regardless. Once the burrow is as deep as its body length the fish changes its direction and starts travelling back to the surface, creating a U shaped tunnel in the process.

While the tunnel remains moist the lungfish travels continually to the top of its burrow to breathe. But soon even its deep burrow dries out. All activity now comes to a standstill. The lungfish begins to aestivate, the fish version of hibernation, resting in a slumber that will remain until the rains finally return. Its mucus hardens into a muddy cocoon around it, keeping the animal's body moist even as the soil itself dries out and blows away. At the top of the cocoon is a small hole that lets air into the mouth, so that even as the fish slumbers its lungs remain breathing and maintaining its body's vital functions. As time passes the basic processes of life almost come to a complete stop for the fish, its blood pressure drops, its heart barely beats and it uses less oxygen. Any fat that is stored inside its body slowly burns up; it is the lungfish's sole source of energy until the rain returns. Once it has used up all its fat reserves it will start burning muscle and once that is exhausted its basic life systems will start to become affected. If the lungfish has eaten well during the wet season than it can remain in aestivation for several months or even longer. But occasionally the water will fail for several years. In these situations many of the juveniles and the weak will likely perish. But some healthy West African lungfish can even survive these conditions and continue waiting. And eventually the rains will come.

After so long without rain what results is often a flash flood. Cracks in the earth result in great grooves for the water to rush into. As the soil takes on moisture it begins to expand again. As the deluge rushes across the desiccated lands the work of eons of lungfish excavations create runways that drive the water towards the burrows. Rapidly the hole fills and the hardened soil around the cocoon begins to wash away. As the water soaks into the fish's skin it begins to wake up and resume its bodily functions. Almost immediately it begins to drown. The water that replenished it is also suffocating it and the struggling fish rushes to escape from its cocoon and return to the surface. It gasps, filling its lungs with air before it can recover enough to take advantage of the wet season. Now that the rain has returned the West African lungfish will, once more wriggle across the endless swamps, taking advantage of the expanding wetlands that few other fish will have been able to colonise.

Other African lungfish typically live in areas moist enough that they do not have to cope with such severe conditions. But in captive situations all African species can produce a cocoon and commence aestivation until moist again. Even the spotted lungfish from the lush swamps of Zaire can endure artificial drying by building a protective cocoon. But the fins so characteristic of the group are not designed to support their bodies out of water; they are so fragile and elongated that they seem almost an encumbrance when it dries. That is because the modification of the fins of both the African and American species is a comparatively recent development.

Lungfish belong to a group called the lobe finned fish that are almost as distantly related to other bony fish as the cartilaginous sharks are. Long before the sturgeons started appearing in the Permian and millions of years before the tarpon started terrorising the oceans a group of bony fish took a different path in life. The lobe finned fish group separated from the more common ray finned fish group 440 million years ago and the major difference between the two is indeed their fins. Rather than thin struts more ancient lungfish had solid muscles to control their fins. Like the arrangement of our own muscles with our biceps and triceps, having the muscles directly connected to the fins made their movement more precise and powerful than their ray finned competitors. But this specialism also made lobe finned fish less flexible swimmers compared to other bony fish and they died out in the oceans, and indeed elsewhere, almost completely. Such solid fins are barely recognisable in the existing African lungfish let alone the American species. For hundreds of years the most accurate reconstructions of primitive lungfish fins were based on a creature known as Ceratodus. While the scientific studies of this fish were thorough they were limited when it came to understanding how it actually used its extraordinary fins, let alone its behaviour. Further explanation was unlikely as Ceratodus had been extinct for some one hundred million years.

That was until unusual reports came from Queensland, Australia, in 1870. Ceratodus had been rediscovered. More detailed observation of the skeleton showed that the new fish was not identical but it was scarcely changed from the original species. And it clearly demonstrates what was likely the original purpose of those muscular fins. It uses them heavily like paddles, in a manner that could scarcely be called fluent. Certainly, the Queensland lungfish does not move at any great speed and it would be hardy be noticeable but for its size and the fact that it regularly rises to the surface to take a loud gasping breath.

Queensland Lungfish, North-west of Brisbane, Australia- ***Tse Chung Yi***

The Queensland lungfish has just one lung and although it often surfaces it relies on its gills more for its respiration needs. Unlike its relatives if its pond dries up it will perish. It is unable to make a cocoon; at best it can shelter beneath a pile of damp leaves while they are still moist enough to support its gills. When the pond is full the lungfish glides above the soil, seeking out hidden tadpoles and worms. It uses tiny electrical receptors under its chin to pick up any indicators of life. These receptors are only effective when the lungfish is directly above its prey. So the lungfish drags itself across the ground with its solid flippers in a movement that resembles terrestrial reptiles more than that of a typical fish. The fins that produce its peculiar walk are very similar in shape to those of the first amphibians' limbs.

There is one other fish that gives an even closer vision, if possible, of the fins and behaviour shared by both the lobe finned fish and land vertebrates. Around 425 million years ago the lungfish separated genetically from a group of relatives with equally sturdy fins. Unlike the lungfish these were an entirely marine group and highly successful at the time, the coelacanths. Many had highly developed lungs, as their distant freshwater relatives did. More than 120 coelacanth species have been identified; some living in the deep sea, others gathering in largest numbers around the coastline. But none were found more recently than the end of the Cretaceous, 66 million years ago. The group appeared to have been victims of the same mass extinction event that also claimed the dinosaurs and ammonites. That was before the rediscovery of the Coelacanth.

In 1938 a rather unusual catch was brought in off the coast of Africa; an ancient fish that while not a shark was clearly different from any other bony fish. Like the Australian lungfish this was a living fossil, almost identical to its extinct pre-identified ancestors, although the survivor appeared to lack any remnant of lungs. Further specimens were found in Madagascar, and the Comoros Islands where they appear to be most common. Rather than the group being completely extinct a second smaller species was discovered sixty years later in Indonesia. Molecular studies show that the two species split from each other about five million years ago. Evidently these ancient fish have changed more than their appearance alone may otherwise indicate.

The coelacanth's habitat reveals part of the secret for its long absence, both to mankind's knowledge and the fossil record. It inhabits some of the deepest parts of the ocean, areas where the main source of food is debris and bizarrely planktonic droppings. Here food for large predators is incredibly sparse and the coelacanth may have to wait several weeks between meals. These are nocturnal hunters and during the day they hide in deep sea caves. Generally they enter their own cave but occasionally groups of eight or so may share until they leave to hunt during the night. Squid are common in the darkness and this ancient fish can grab them in short pursuits. But it has found another way to find prey. The coelacanth performs summersaults, bringing the electrical sensors in its head above any hidden prey. Like the lungfish its solid limbs are essential in finding a meal. The coelacanth uses its solid fins to grasp the seafloor, holding itself upside down for a few minutes. In this position the coelacanth can detect the movement of any food below it.

That impressive head was enough to catch the attention of scientists in the 1930's, even before people had been lucky enough to observe it in the wild; newer technology and techniques have now turned some of the original impressions on their head. The muscles around the head, particularly around the jaws, had been misidentified- they weren't muscles they were ligaments so the coelacanth's mouth is incapable of puckering up to suck in prey like ray finned fish do. The jaws are even more remarkable, they primarily consist of cartilage- a characteristic more in line with sharks, rays and chimaeras, than other bony fish. Having such high levels of cartilage imbedded in the jaws provides equally high levels of pliability, allowing a hungry fish to suddenly expand its jaws outwards to grab its unsuspecting prey. Genetic analysis, something else that wasn't available in the 1930's, confirms that despite this anatomic abnormality, the coelacanth, and also the lungfish, are more closely related to other bony fish than they are to sharks. The re-examination of the skeleton also found something else: the missing lung. Buried beneath a heavy layer of muscle and tissue it has become an obsolete organ, and would be of little use to the fish in the exceptionally unlikely event of one attempting to breath air (if it was even capable of surviving the substantial change in depth). The redundant lung indicates a possible reason for the modern species survival, and how it was able to survive for tens of millions of years without leaving any traces in the fossil record.

Having active lungs restricted the other coelacanth species to the top layers of the sea- allowing them an advantage against their competitors by simply popping their head above the water any time they needed to rapidly raise their oxygen levels but also restricting them from

African Coelacanth, 68m depth, West Coast of South Africa- *Bruce Henderson- WrecklessMarine*

moving to deeper, and more stable conditions when they had to. On land the lungfish were able to survive the catastrophic end of the Cretaceous by sheltering within the accumulated mud and debris until the worse of conditions were over. Fresh water fish survived the event comparatively unscathed; perhaps in eggs, or in deeper lakes where they could seek shelter and scavenge until conditions improved. Those fish that lived in the oceans would be less lucky.

A combination of increased temperatures, the resulting acidification of the seas and current changes, pushed many marine species past the limits of survival. The primary reef forming invertebrates were obliterated, as were the ammonites, many species of sharks, and virtually all coelacanths. Those coelacanths that, like the current species, had lungs that were more defective than not, could continue living in the depths- lives virtually unthreatened by any catastrophe above. Any individuals that were lucky enough to have been fossilised are likely deposited too deep to be practically extracted and returned to the surface. But above them, as conditions improved, the surviving coelacanths proved unable to adapt quickly enough to reconquer the seas- that would fall to corals, which would later form reefs capable of rivalling any of their fallen competitors, the ray finned fish, and finally the surviving sharks. For coelacanths the gradual increase in the variety of prey species returning to their habitat would have been the only indication of recovery in the rest of the world, although such a change would have occurred too gradually for it to have been noticeable in any individual's life time.

Further investigation has revealed another secret that the coelacanth had been keeping. Inside the female's body can be from twenty up to sixty eggs. Just like sharks coelacanths fertilise their eggs internally. Not all of them will develop but those that do can expect additional special treatment. Even after hatching the young will remain inside their mother. A substantial amount of yolk sustains the growing infant, allowing it to rest protected inside its mother until it reaches an advanced stage of development. There is no placenta or other mechanism for the female to transfer nutrients to her young; instead they have to rely on their large yolk sacs to fuel their growth. When the female gives birth only a portion of the original young will have survived. A dozen is a good sized brood for most females. Even the largest females give birth to no more than thirty young at a time. But these young are nearly forty centimetres long and basically miniature adults. Those that survive adolescence may live for more than sixty years. Risks are high though. It is estimated that the largest population of African coelacanths, situated around the Comoros islands, would be 450 at the most and more pessimistically, if unfortunately realistically, perhaps as few as two hundred or so individuals. The number of Indonesian coelacanths is unknown, estimates are perhaps ten thousand mature adults- but against that there have only been ten confirmed sightings, including some unfortunate individuals trapped within fishing nets- even though they are legally protected.

The coelacanth's unusual method of child birth is of particular interest to us because it is reminiscent of the same nurturing techniques employed by most mammals. The coelacanth, as is shown clearly by the formation of its fins, is one of the closest aquatic relatives of land based backboned animals but it has discovered live birth independently from the mammals. For many years it was uncertain whether the coelacanths or the lungfish are the closest relatives to land vertebrates but genetic comparisons have removed some of this confusion. These two groups of fish may appear closely related, as indeed they are, but the lungfish is actually more closely related to animals that spend most of their lives on land than to any marine fish, including the mysterious coelacanth.

Roughly four hundred million years ago some enterprising members of the lungfish group began to take advantage of their breathing techniques to venture further on land. Initially this would have been only a temporary environment but those individuals that were born with small advantages like stouter limbs or improved breathing would have used it increasingly. Four hundred million years is an immense amount of time to try to comprehend when a person who reaches a lifespan of a hundred is seen as extraordinary. During a century of life, if we are lucky, we grow, learn what to eat, how to remain safe and fight to raise a family to continue such a process after us. These are not new concerns and the original land lubbers survived by stumbling upon various strategies successful enough to be copied by their offspring. Just as with even earlier species, such as Pikaia or Haikouichthys, some fossils have survived to the modern day to give some impression of these early innovators, like the Tiktaalik, a fish with limbs sufficient enough to support its body at least partially out of water and a neck, reminiscent of the salamanderfish, that enabled it to spot new opportunities and dangers.

Reconstruction of Tiktaalik, one of the first vertebrates able to venture on land- *Zina Deretsky- National Science Foundation (USA)*

The earliest surviving representatives of this process, the amphibians, remain indebted to the fresh water of their ancestors by the necessity of keeping their skin and offspring moist, indeed several of them produce wriggling larvae virtually identical to a juvenile lungfish. Other descendants of these original pioneers developed water proof skin and hard shelled eggs and conquered the deserts of the world; hunting down insects and each other with the passion that had initially arose in the ocean.

Just as in the water the initial movement of these animals was controlled by the surrounding temperature. Tuna and mackerel sharks had found a way to thwart such restrictions by fuelling their own body heat internally and so did two groups of reptiles. One, the theropod dinosaurs, insulated themselves with fluffy scales, feathers, that later enabled their direct descendants, the birds, to soar into the sky with even more skill than a hatchet fish. The second group of warm blooded reptiles also relied on modified scales, hair, to keep warm and these mammals eventually would come to dominate the ground just as effectively as the birds did the air above them. The effect of both of these initially land-based animals was enough to transform the environment, affecting and allowing the creation of grasslands and rich rainforests just as the combined effect of a multitude of fish in a reef allows the whole ecosystem to thrive.

So immense is this diversity of land vertebrates that at first sight it is hard to recognise that they are related to each other let alone fish that shoal in the sea. But the reality is that the demands of living in different habitats has changed the shape of their inhabitants; something that is clearly visible when a dolphin and tuna are observed swimming beside each other- even though the dolphin's earliest ancestors lived on land. If we were to apply the same rigours of classification to vertebrates on the land as we do to those in the sea it would be fair to say that all terrestrial backboned animals are, in effect, land fish. Indeed there is a smaller genetic difference between say a tiger and a coelacanth or even a salmon and a turkey than between any of these animals and a shark, and all of *those* species are more closely related to each other than a hagfish. This kinship is most visible when comparing embryos; even that of a human develops gills while in the womb, before such structures later change into jaws and other decisive parts of the anatomy.

Just as mammals, coelacanths and great white sharks have independently discovered the benefits of live birth, some species share other traits, such as a desire to control a food source. Along the banks of the Yangtze five thousand years ago people started to harvest the rich food that was available. In the river were massive paddlefish and catfish but there was something else growing at its edge, a type of grass with grains that were unusually easy to gather. Rice. Such a crop required immersion in water in order to thrive but the plants would often be naturally separated by vast distances; due to the impact of travelling herbivores and competing foliage. Like the western buffalo bream of Rottnest Island these people began to deliberately eradicate the plants they could not consume and drive away competitors, but their flexible hands meant that they could also drag up and grow rice together to make them more accessible. Unlike the territorial fish these individuals were highly social which allowed them to spread their culture and the technique of producing large paddies for easy harvesting. The spread of rice and urbanisation that went hand in hand with each other drastically changed much of China and was the first case of what would soon occur elsewhere as the human population increased. It began to produce the first significant threats to many species; and the growth of urbanisation has also lead to an increase in the rate of extinction during the last couple of centuries. But not all fish were adversely affected by such changes, for meat was just as desirable a meal as rice and controlled just as efficiently.

One fish thrived in the rice paddies, easily finding enough food and reproducing, carp. They were desirable not just for their flesh but for the waste with which they readily fertilized their environment. So carp were spread across Asia while simultaneously, in Europe, other communities were finding the benefit of introducing carp into new areas as a ready source of food. The key development of carp didn't really occur until a thousand years ago when they were introduced from China into Japan. For several hundred years they spread around the four main islands until during the 1800s farmers looking at their properties for new sources of trade began selectively breeding the characteristics that had occurred during isolation. The result was the koi, a fish shaped by man to be seen from above. The first feature to be developed was a bright red cap on the forehead, only for other patterns and colours to become favoured as the fish trade continued. Early in the Twentieth Century a shoal were introduced into the Emperor's Imperial Palace, but their introduction into the Western World, first into the United States and later into Britain, only really occurred after the Great Depression. Since then the koi population has expanded well beyond the constrictions of its native environment.

The popularity of koi has meant that in less than two centuries there are now more than a dozen separate groups of this carp recognised. One of the key breeding pressures on koi is that they were traditionally viewed in ponds, so only those with appealing patterns on their back were likely to be bred. Any fish with curious patterns or desirable colours on its underside would not be recognised by traditional breeders and would consequently be culled out of the gene pool. But another smaller species of carp was also selectively bred for its appearance, this time not in Japan but in its native China, the goldfish.

Two thousand years ago there was just one kind of goldfish, with the few regional differences common to a species. But by 300AD they were commonly bred for food, until, like the koi, after another couple of centuries they came to the attention of the imperial family. The fish began to be specially bred for the amusement of the nobility, resulting in the selection of unusual features like bubble eyes and elaborate fins which would have resulted in their possessor being outcompeted in the wild.

Domestic goldfish- *Sommai*

In the Nineteenth Century they were some of the first fish from the Far East introduced to the West; capable of living even in small containers with little food or air.

Such adaptability brought the group an unfortunate reputation of stupidity, one which can soon be dispelled by the keen observer who witnesses the fish maintaining strict social relations including bullying, regularly coming to the surface at feeding time (even before the owner has picked up the food), or swimming through a hoop after training. Some individuals have even been able to be taught how to head-butt a miniature soccer ball into a net, responding rather rapidly to food rewards being provided as they learn how to do this. Such training would not be possible if goldfish had a mere three second memory, as is commonly paraded as a fact. Part of the reason for this myth is that goldfish lack a hippocampus, the region of the brain that in mammals is associated with memory recall, yet the fact that they can learn suggests that they must be remembering things somewhere else in the brain.

Scientists who tested goldfish in mazes found that by the second day they were already proceeding quicker through the maze- swimming directly towards the food. This trend continued to accelerate over the next four weeks; dropping from taking an initial average of twelve and half minutes to find the food to an average of a little over 17 seconds.

Other experimenters have found that goldfish can recall the escape path out of a net some eleven months after last attempting it- a combination of navigation and memory skills that many humans would struggle with. So capable are the goldfish at adapting that those released as unwanted pets have now spread over six continents, despite originally only inhabiting a limited part of Asia. The effect of people on this creature's diversity is clear. Since the time of Christ a single species of goldfish has been modified into nearly twenty different kinds.

Other novelty fish were soon spread around the world. The walking catfish of Asia proved a temporary curiosity for a number of home aquariums in the United States of America for they soon began appearing in wetlands across the country. Being hunters they immediately began to target the tempting variety of prey. Nowhere was the presence of the catfish more severe than in Florida. Here was the home to the nation's pristine wetland; the Everglades. Such an unspoilt environment was also ideal for the walking catfish. In desperation people turned to extreme measures. They began poisoning small ponds known to harbour the invading catfish. They had reasoned against the adaptability of the once popular pet. With their sensitive barbells and taste buds running across their bodies the catfish could detect the smallest trace of poison. Then they turned to their unique ability to survive. They simply abandoned the toxic water and walked to the next available waterhole. By the time the poison level dropped in the original pond to a tolerable level all of the catfish's competitors were dead. In the predator free pond populations of aquatic invertebrates rapidly increased, creating a fertile feeding ground for the next walking catfish to arrive.

Feral fish are becoming a problem worldwide and not all introductions are accidental. In Africa the isolation of large lakes for millions of years had led to the diversity of one group of fish, cichlids. In Lake Victoria alone were, depending on how rigidly you classify a species, more than 450 different types. This in turn led to small fishing communities spread out across the lake. There was however no large trophy fish so one more fish was introduced, the Nile perch. Fossils showed that Nile perch were previously native to the lake, prior to the huge radiation of species. What was not as well understood was how much of the evolution of the cichlids depended on the perch' absence. Within fifty years of introduction fifty species of cichlid, nearly a fifth of the lake's species had been driven to extinction. Another 130 species are critically endangered, meaning that they are only a few precious individuals away from disappearing completely. The perch did not drive those species to extinction by itself. In addition to the new hunter other pressures had been placed on the lake, in particular increasing levels of pollution and decreasing levels of oxygen- forcing the struggling cichlids into more oxygen rich, and exposed, waters creating the ideal opportunity for the perch.

In Australia a close relative of the Nile perch, the barramundi, has also been spread widely through the continent. Again it is a 'native' fish but previously its distribution was restricted to rivers it could reach, either from the sea or during flooding. Its significant size made it an ideal target for fishermen after European settlement and it was consequently released en masse in basically any significant body of water that could be reached. The barramundi, like the Nile perch, is a top predator with a population usually limited by the amount of suitable prey. But many barramundi are now being raised in fish farms and released as adolescents into new river systems. Many will eventually be tempted by worm and hook but in the meantime they survive by eating their way through the food chain until there is not enough food left to support even hatchling barramundi. Other predatory fish are also being transported around the country; increasingly spangled perch, relatives of the barramundi that can survive in streams too constrictive for other hunters. Even the ancient saratoga has been moved around, exploring waterways where it has not been seen for millions of years. With their superior parent minding the saratoga are able to make the best of their new environment. But other fish are less fortunate. The delicate rainbow fish that flit around in courtship of each other are quickly snatched up and are vanishing right across the top end. A few species survive in just a single waterhole or feral pig wallow. A drought could knock them out in one go but equally so could a flood by finally allowing these transported predators access to their oasis.

There is another, less obvious, cause for concern with such an exchange of fish. Isolation has created different populations of fish, not just rainbow fish but also perch and barramundi. These separated groups are too similar to be formally recognised as different species or even subspecies, yet, but variations whether in eye colour, scale shape or behaviour are clearly there. Via mixing these different communities together these unique characteristics are either spread through the entire population or overwhelmed. Either way the fish remaining have lost some of their unique quality and their genetic variety. But sometimes there is no option but to transplant fish. Saratoga in the Burnett River not only fed on peaceful shoaling fish they also impinged on another great hunter, the Queensland lungfish. For the saratoga's unwilling competitor conditions are not quite as favourable. The lungfish only naturally inhabited two river systems and its slow birth rate put it at substantial risk of disappearance. But even it has been introduced into some of the few rivers it lives in today, including the Fitzroy where in the latter, ironically, it competes with the saratoga in *its* natural waters. This may have ensured the lungfish's survival but it has also placed other native fish like gobies, rainbow fish and hardheads at risk of increased predation and extinction. To confuse things further in one of the rivers where lungfish were considered to be introduced, the Brisbane River, further genetic and fossil have confirmed that the lungfish there are a distinct population- so unlikely to have been transplanted there, although some individuals might have been. There are no simple solutions in the management of fish, particularly when both prey and predator are at risk.

In Europe, for one species, things have become, if possible, even more confusing. A Romanian fish, the Techirghiol stickleback, speedy territorial fighters like its relatives elsewhere, has likely been hybridized into extinction by an introduced closely species from America. Ironically enough some of the descendants of the extinct population likely still live today in Lake Techirghiol, but due to the invasive status of the American ancestors that contributed some of their genetic makeup, they are granted no legal protection today.

Ziebell's Handfish, cave near Eaglehawk Neck, Tasmania- *Andrew Green*

One of the unique challenges of conserving fish is the great fecundity of many species- when just a single individual is capable of producing several thousand eggs at one time it can be difficult to identify when a species is under unsustainable pressure, and act in time to protect it. This puts many previously successful species at risk of extinction. For the Chinese Paddlefish such an event has likely occurred. So too for several of the native fish of the Great Lakes of Canada- for while some were impacted by invading lamprey other species were already on the very brink from over-fishing before the parasites had even reached the lakes.

The seas themselves were considered reasonably immune to extinction. That changed in the early 2000's with the announcement of the extinction of a species of ray, the last time it was seen alive Charles Darwin was still alive. In Tasmania a member of one of the more charismatic fish families has potentially joined it, a handfish. Ziebell's Handfish, was so adapt at walking that to see it moving it would be easy to initially mistake it for some bizarre type of marine salamander- before realizing that no such creature exists. Like the rest of its family it would use its angler, gently waving it around until it caught the attention of unwary prey.

For most of the year Ziebell's Handfish was said to live deep within the caves but winter would trigger them to start their slow amble towards the entrance of the cave in the hope of attracting not food but a potential mate; although sightings were rare enough that uncertainty remains regarding practically all of this little critter's behaviour. The movement that so characterised this handfish and other members of its family wasn't particularly fast, certainly not compared to the tiny feeder fish skiting through the same water, nor would its walk enable it to leave for more suitable environments. And *that* would prove its vulnerability.

Pollution and environmental damage began to take its toll. In the 1990's, when the striking purple individual in the photograph was observed, it was already clear that it was at risk. The last time it was sighted was likely during 2007, despite its habitat being centred around Hobart, Tasmania's capital. There is a very high likelihood of it being extinct. One of its relatives, the smooth handfish has already been declared extinct in 2020, although that was reversed a year later as there was not considered to be enough data to confirm whether this fish was still alive or not. That species found its habitat destroyed by the dredging involved in intensive oyster harvesting in the last and the 19th century. Any smooth handfish able to survive the destruction unleashed by the dredging would have found themselves exceptionally vulnerable to the appetite of larger predatory fish. Regardless of whether these two handfish species are completely extinct or, perhaps, survive in populations so small that no reporting has occurred for decades, it is a rather unfortunate fate for such charismatic and unique creatures.

Nearly all fish, ultimately, are prey for one particularly enterprising omnivore, man. That is not a particularly new development. Already by forty thousand years ago, people were eating fish and in such high quantities that the chemical analysis of the skeleton of a man found near Beijing, known as Tianyuan man, revealed that he regularly ate freshwater fish.

In East Timor, at around the same time, people were also relying on fish for their sustenance but by twenty thousand years ago they made a major development, some of the first fish hooks. They were not content to feed on just freshwater fish here, not with the considerable temptation provided by the smorgasbord available in the surrounding ocean. Fish bones from more than two thousand eight hundred different individuals show that about half of the species targeted were not just marine fish but those at home in the deep ocean. To reach the tuna, deep sea sharks and rays that their community now relied on required advanced maritime craft capable of venturing out well beyond sight of the shore.

People in other parts of the world, independently, also came to rely, and continue to rely on those deep sea species, but increased levels of industrialisation has placed a level of pressure that even the most fecund of species can struggle to survive. The decimation of the simultaneously economically and ecologically important cod is the most famous of these cases but sadly is not alone. Indeed, such mistakes are being repeated worldwide. In Europe strong fishing pressure has produced remarkable changes among skates, the most common rays in this area. The pressure exerted has largely been due to the size of the holes in the fishing nets, which are strictly regulated. This strict regulation means that, particularly in Britain, individuals of smaller species are able to slip out while larger kinds are unable to escape. This unforeseen selection process is producing a substantial increase in the population of smaller skates as their competitors disappear and, due to their diminished size, such individuals are less likely to be caught and consequently less valuable to local fishermen.

Bigger fish are also being targeted in Eastern waters, particularly tuna with the spread of sushi throughout the world. As their numbers deplete the remaining bluefin tuna have become increasingly expensive and subsequently a status symbol. In Japan, where they are most heavily prized, one prize specimen sold for more than three million American dollars. It appears that the creation of a cultural and status symbol may rob people and the oceans of the most magnificent of the trophy fish in just a few decades.

And it appears we are poisoning other fish that we could prey on too. Chemical run off into the oceans is inadvertently absolved by the microscopic plankton and algae that are the foundation of the food chain. Those that feed on the plankton directly accumulate the poison which they are unable to expel, but its mild concentration has little effect on them. Any hunter that consumes them however digests the same chemicals and is likely to eat several affected prey. Higher up the food chain, such as in barracuda and billfish, these chemicals build up to such levels that it affects reproduction. It is like the spate of DDT and other pesticides last century that began to reduce fertility rates of birds of prey on land. But while people do not typically eat eagles they certainly eat fish that are at the top of the food chain and are consuming the same chemicals which built up for several years in the hunter's body. The result, ciguatera poisoning, can lead to miscarriage or even the fatality of an adult where consumption is high. So now humanity with our lacklustre use of pesticides and other poisons has begun to poison our own communities.

There are other fish with such an undesirable taste that for many years they were kept off the menu. To increase their buoyancy sharks have high concentrations of uric acid around their body which also reduces their palatability. But, just as with the first unpleasant flavour of caviar or fugu, the initial foul taste is soon ignored, especially when a meal becomes associated with status. In China for hundreds of years the nobility enjoyed the privilege and associated slaughter of shark fin soup. With the rise of communism this fancy meal was brought to an end. This brought only brief relief to the shark population. The release of the book and movie Jaws turned the shark from a maligned animal into a prize trophy. Sharks had previously been hunted but that was nothing compared to the global mania now desiring their skin. In a time when having a rhino head stuck on a wall was seen as barbaric the head of the great white began to replace it. But the harvesting of these animals was not limited to the great white. To the public one of the most awe inspiring catches was that of the grey nurse shark with its mouth full of teeth projecting out its jaws. The grey nurse's diet was restricted by the very jaws that made it appear so fierce-some. It could eat little more than fish; even blubber coated seals could not interest it. This was no man eater but the sight of it being dragged out of the water wowed the crowd like no other sight. Nowhere was the hunting more severe than in Australia. By 1985 the population had dropped so low that it was declared critically endangered and became legally protected in Australia. It was the first time this had happened anywhere in the world. This decisive action saved the species but other sharks could not be so easily protected.

Further out at sea the oceanic white tip was once considered the most numerous shark, indeed the most numerous large predator on the planet. When a whale was harpooned these sharks would rapidly appear in groups beside the carcass, helping themselves to the flesh while its bloated body was dragged towards the ship. The whitetips were also significant bycatch. What was not considered was how far the oceanic white tips had travelled to reach the meal, led by their keen noses to the scent of blood in the water. The dying whale had attracted sharks from nearly 130 kilometres away. The population had been significantly overestimated. Even before considering the impact of bycatch sharks were already on the edge of a precipice and it was about to collapse.

Simultaneous to the increased horror movie treatment of sharks another species was receiving an epic media makeover. A little over forty years before whales had been considered that monstrous that one even featured in the Disney children's movie Pinocchio swallowing a man, child, their wooden raft, pet cat, and ironically, a goldfish, and only releasing them when its throat was set alight. Adults keen on thrills and popcorn could see the possessed sperm whale Moby Dick sink a ship and furiously pursue the crew, while reminding themselves it was loosely based on a true story. But increased scientific, and most importantly, public interest and empathy in these remarkable, yet still dangerous, animals was changing things, even as they became increasingly rarer, and rarer. In 1986 an international ban on whaling came into force. There was still a need for a rare meal from the sea fit for the elite. So the next year China allowed the trade of sharkfin soup to recommence. This time the meal did not remain restricted to royalty, its only restriction was its cost. This status symbol also became exported to other countries which converged in harvesting the sharks in great numbers. Between seventy to a hundred million sharks are estimated to be killed each year for their fins. Some of those threatened sharks like the great white are protected in particular countries like Australia, New Zealand and South Africa. But such isolated protection has little effect on fish that migrate between the oceanic borders of so many countries. Only a worldwide approach will guarantee the existence of such important creatures in the future.

Death Valley Pupfish courting, Death Valley National Park, California- ***Floyd. E. Hayes***

In some places people have banded together to ensure the survival of a species. In California is found one of the most extreme environments for any fish, a desert so severe that it is called the Death Valley. Yet in a few secluded spots are isolated oases where fish have been able to colonise, one group in particular, the highly adaptable killifish. Here is found one of the hardiest of the group, the Death Valley pupfish. It has to be tough because even these aquatic havens can still heat up to 40°C each day, enough to seriously damage the internal organs of most fish. The associated evaporation has other effects. Over time it leads to an increase in salt creating, in effect, an inland sea. The pupfish are sturdy fish but even so many of them do not make it through their first year.

Water in a desert has value to more than fish; domestic use, agriculture and, most aggressively of all, mining all began to suck up it relentlessly as industrialism began encroaching deeper into the desert. The water table began to dip and the future for the pupfish looked grim.

Small and unusual as it was this fish somehow managed to capture the hearts and attention of local people, it was like the Californian version of a panda. It became the state's focus on what would be lost if people continued to wreak damage incessantly on the natural environment. The community managed to pull together, not without conflict, resulting in specific rules being made and enforced to ensure the survival of the rare killifish. The most fitting testament to the success of this is that the Death Valley pupfish, unlike the Chinese paddlefish, many African cichlids and possibly some of Tasmania's handfish, is not extinct and still exists despite its perilous choice of home in the desert. The most effective means of protecting it in this case was not just protecting its limited habitat but encouraging it to breed in captivity. And breed it certainly has. The initial population of about five hundred Death Valley wild pupfish has raised dramatically thanks in part to the continual level of human protection and support surrounding them, including the release of over a million of its captive relatives into the wild. This is just one example of the positive interactions that can and are occurring between people and fish.

Back out at sea among the most critically threatened habitats are coral reefs; very few have been able to escape the impact of bleaching events in recent years (where the algae that give the coral its colour, and feed them via photosynthesis, are expelled when the water gets too warm), and the increased acidification of the seas- accelerated by global warming- is making it increasingly difficult for them to repair themselves. But even here there can be, and are, positive impacts on fish, sometimes in surprising places. Abu Dhabi, for instance, was ranked exceptionally low for sustainable fishing- even its own environmental agency recorded it as having not even 9% sustainability as recently as 2018. There was a push for change; different fish traps were banned, further regulations were introduced to regulate commercial and recreational fishing, and they worked with the community.

The changes were more than just on paper; they also included having six national parks and, most impressively, planting artificial coral reef colonies; from 2021 till 2025 a million coral colonies had already been restored. And it is slowly beginning to work. Small fast breeding species, like clownfish and surgeonfish, were able to take advantage of the change comparatively quickly, but now other larger fish are beginning to appear; oceanic triggerfish, the pinjalo snapper (a species that had been particularly hard hit by earlier unsustainable fishing), and most impressively the first sighting of a white-spotted grouper, a large top predator species, suggesting that nature is finally getting closer back to normal- at least here. The sustainable fishing index rating was now more than 97%, and clearly the local fish were enjoying some of the improvements that had been applied to improve it.

Stranded Basking Shark, Styrsö, Västra Götaland, Sweden-*Björn Källström*

Artificial reefs are not restricted to Abu Dhabi; other countries around the world, including Fiji, Indonesia, Japan, France, Thailand, and in the United States of America the very largest- Wheeler North Reef, in California, at more than 370 acres, are embracing this technology to extend remaining habitat and reduce existing environmental pressure.

In Mo'orea, French Polynesia, then teenager, Titouan Bernicot, started up an organisation, Coral Gardeners, establishing and maintaining artificial reefs. Here too fish have taken the lifeline provided to them and moved into their new abodes- visibly establishing populations in the new habitats within a month. It is proof that even in increasingly human influenced environments fish can find a place for themselves, and even thrive, if they are given the opportunity to do so.

And even the most unlikely of species can attract the attention of those willing to help. In September 2020 a rather unexpected creature was found trapped in a shallow bay on the Swedish island of Styrsö. At least six metres long the first observers could have been forgiven for mistaking it for a small whale. But no, this was something rather different, a Basking Shark. Despite its significant size it was still growing and only half size. Perhaps it was its inexperience, or perhaps poor health, that lead to its fins being firmly within the bay's mud. What was clearer was that the young giant was struggling and would be incapable of returning to the seas using its own strength alone.

A team of rescuers, including marine biologist, Björn Källström, and the local Coast Guard, worked over many hours to free the animal. Initially they attempted to push the shark back into depths deep enough for it to be able to swim away, but with that proving ineffective they instead wrapped a matt around its large back fin and then tied the connected fin to a boat and towed it out. In its confused condition the shark soon returned back into the shallow waters and began to strand again. It would take until evening when the basking shark gradually began to settle until they were able to finally return the unlucky fish back into deeper waters, allowing it to finally swim freely again.

Unfortunately, several days later the unlucky giant was washed up dead, the stress and inability to breathe adequately had likely proved too much. Despite that, the fact that so many people were willing to devote their time to a creature about as different from us as is possible shows that change and compassion is possible- even for the most unlikely. Should a basking shark be unlucky enough to strand in Sweden in the future those who find it will do the best they can for these unexpectedly charismatic giants.

Other people elsewhere are already having more success. On India's Western coast, twenty years of community and scientific support has worked to help protect the whale sharks there, including compensating fishermen whose nets are damaged by entangled leviathans. Just one team alone, in Kerala, has released fifty of these magnificent fish during the past seven years. Equally importantly, and one of the primary reasons for the newfound success, the local communities- those who rely on the sea for their sustenance, income, and as part of their day to day life- are on board and actively part of ensuring that their special aquatic neighbours are looked after.

Fish may appear at first sight only simple creatures, merely a primitive reflection of our most distant ancestors and nothing more. Such an opinion is clearly an illusion the further one looks into the varying behaviour of the group. Fish and their relatives are the most distantly related of all backboned animals. They are also the most numerous and diverse group, including highly modern creatures that hunt with electricity, are able to feed using just their skin, create their own light in pitch darkness and some even dare to fly. The ancient relatives of the fish; tunicates like the salps are also showing themselves as adaptable in a constantly changing world. Many of these fish seem odd but perhaps their oddest moments are when they remind us of ourselves. For millions of years they have raised families, avoided danger and enjoyed a good meal where they can. Were it not for their aquatic nature one could say that they were almost human.

▼Leafy seadragon- ***Dmitry Rukhlenko***

Afterword and Acknowledgements

Wow.

That is perhaps not exactly the first word you were expecting to read at the end of this book. But, yep, wow. But this book had an unusual triggering point so perhaps wow is a fitting way to conclude it.

So, to really simplify things, David Attenborough released a book on reptiles yearrrs ago, and on the back of the blurb stated that it completed his series on all of the major animal groups, and I thought 'he hasn't done one on fish' so I decided to write a book for him. That took a Bit longer than I thought; then when it was all finished... several years after initially starting writing it, Mum encouraged me to send the book to the British author and naturalist. So, I googled his address and sent him an envelope with a USB of the completed book, with the English nature writer listed as the main writer which seemed only right as the book was for him. I later received a letter back from him stating that he didn't have a computer, then a bit of time passed and with more encouragement from Mum we went to the local printing shop and printed it out (following another edit) and sent it to him. This time I was lucky enough to get another letter back from David Attenborough thanking me for writing the book for him and that he was looking forward to reading it.

That was back in 2016. And that would have probably been the end of the book's story as well if it hadn't been for the curiosity and encouragement of friends and family. My best friend Adam, for instance, kept asking me about this book that I'd written and if I was going to do any more with it. So, after nine years, I thought it was time to do an update, a complete re-edit, and release it for a general audience.

The trouble with science, from a re-editing point of view, is that it is an industry where the workers are never happier than if they could prove the facts of previous generations wrong. A discovery that is identical to a previous discovery isn't a discovery- it's a confirmation, and there have been A lot of discoveries made in relation to fish since 2016, even more so compared to 2009 when I first starting writing this. Some of the new science is extraordinary- ultra-sounding whale sharks for instance is something that wasn't even thought-of as a possibility in 2016. Some of the changes have been less easy to embrace from the point of view of them coming with a Lot of work; early fossils that were accurately identified as fish since the 1990s for instance have now been re-identified as not being fish at all- resulting in painful rewrites. Other scientific controversies are looking even less likely to be resolved than they were nine years ago- that revolving around fish gills and exactly what impact they have on a fish's maximum size being amongst the most frustrating one; I have tried to update this work to the most accurate science currently available but I am well aware that by the time you read this there is a chance that the science has changed again. Science likes to do that. Despite my reliance on the work of scientists, researchers, and other nature lovers any errors that are here, and I have tried to avoid any (but I am a realist), are very much my own.

One of the biggest changes, from a personal point of view is now listing myself as the primary author, although the individual who was originally listed as this remains as much as inspiration, and the book remains very much dedicated to David Attenborough.

Even more so, perhaps, than when I wrote this originally it has been a group effort, and certainly with a lot of background support. One thing I do regret in the writing of the original is that I followed the typical format of most BBC books (that I had access to) and didn't reference. That became a pain for two reasons; one when I was updating sometimes I would find a fact that I had written which I couldn't verify anywhere on the internet (presumably due to it coming from a book published well before the internet), but the second reason is that it means you can't acknowledge the work that others have done to uncover the knowledge that you are attempting to share with others. Unfortunately the majority of the works I checked during this time are well beyond the extent of my memory to recall, but amongst the most notable (that I can recall) are: Beer, J. & Hall, D. The Complete Illustrated World Guide to Marine Fish & Sea Creatures; DK. Animal; DK: Prehistoric, Uhlenbroek, C. Animal Life; Readers Digest Australian Encyclopaedia of Australian Wildlife; Byatt, A., Forthergill, A. & Holmes, M. The Blue Planet (also 'Extraordinary Fish' which was also done by the BBC), Low, T. The New Nature (particularly useful for the Australian invasive native species), Summerill, M. Wild Australasia (especially for the western buffalo bream info near Rottnest Island), Alderton, D. Encyclopedia of Aquarium & Pond Fish; Taylor; L. The Nature Company Guides: Sharks & Rays; and (surprisingly enough considering its age- 1966- but this was particularly handy for clarifying details about early attempts to breed tetras in captivity) Innes, W. T, Exotic Aquarium Fishes.

Multiple websites have been useful including Sciencedirect, The Guardian, Wikipedia (yes it has its drawbacks in that anyone can write in it but it also has the advance of links that you can follow to get to the original research), FloridaMuseum, Phys.Org, National Geographic, BBC.com, BES Journals (British Ecological Journals) and FishBase (which maintains active records of all the currently names species and subspecies of fish, regularly updated- exceptionally handy). For those wishing to identify individual species, or to add their own sightings of various wildlife, there is also iNaturalist- and for any lovers of nature it is well worth a look at.

Reviews are a major part of that; reviews sell books. They also make the world of difference to the author- even if someone can't stand the book at least that way we know. Biologist, Daniel B. Lyle, was one of the first to praise the book, and particularly having that from a scientist was absolutely awesome. Other reviewers were kind enough to comment from USA, India, UK.

There are some extraordinary photographers out there, and so many have been willing to include their work in here. You are all such brilliant, remarkable individuals. And when I'm saying all across the planet I'm not kidding. Papua New Guinea, Kazakhstan, Peru, Canada, the Congo, Sweden, Mexico, Greece, Japan, the Caribbean, Alaska, Indonesia, Spain, New Zealand, Philippines, South Africa, New York State and of course Australia, and even Antarctica. That's only Some of the places. Every single one of you made a difference, both to the book, but often with your extra encouragement.

In particular I have contacted so many wonderful people through iNaturalist; all of you are brilliant. Two images of frilled sharks in particular, Lin Ching-Feng's showing a clear view of its trident like teeth, and the other, equally impressive, by Yurimaoh- showing a baby still attached to its egg, were both images that I didn't even think existed *Anywhere* until I saw them and both of you were among the first to approve to including your photos, as was Mandy Gardner with her spectacular photo of an Australian elephantfish. It is very likely that without those first yeses I would have just left with just the small handful of pictures it had in the first edition. A special thanks also to Andrew Green, who not only provided the extraordinary image of Ziebell's Handfish, but also further information about what little is suspected about its behaviour, including its use of caves, with thanks to the dive guide Mick Barron from Eaglehawk Dive for assisting Andrew to locate this species. To Patrycja Lawryniuk who not only provided some brilliant photos to begin with but then dobbed her husband Chris Crippen in, if there were any of his photos that might be helpful- and they were. A special thankyou to all the other photographers who said not only could I use their photos but would I be interested in using more of other species ☺- that was an absolutely brilliant help. An additional thank you to Matthias D'haean who had featured the photo of a mormyrid (elephantnose fish) on iNaturalist and then was kind enough to give me the contact details for African Parks RCA Chinko, so that I could ask for permission to include it- and for Thierry Aebischer for your thanks with this also. Using Fish Base I was able to contact the scientist involved in researching the parasites of the pink vent fish and who took the remarkable photo of one. Harry Rosenthal, as well as kindly providing permission to include some of his photos, provided some additional advice on printing which was really helpful. Specially appreciated this considering he had just completed his involvement in another large fish based project as one of the co-writers of 'Field Guide to Sydney Harbour Fishes'. An impressive book with a lot... of photos and some 700 species, and did I mention those photos- 2,500 of them; the sheer amount of work behind that makes my jaw drop- well worth a squiz at. Two more brief mentions, to Fishipedia (which runs on iNaturalist) and Benoit Chartrer who was able to provide the delightful whiptail catfish image by Xaime Bero- if you have any species that you feel should be included please get involved with this fantastic project. And lucky last- to Björn Källström, with his brilliant photo of a stranded basking shark taken during its rescue which helped to tie up the entire book.

Thank you to Apple, in particular Dee, who helped me struggle over trying to use a non-Apple device to set-up an author account so that I could sell books on Apple. Considering I already had an account for AppleTv and had downloaded iTunes the week before (and could log on to that) you would think it would be relatively easy- urgh- I think we both deserve a very strong coffee after that ordeal, but equally importantly your suggestions eventually worked.

Other people have also been a huge part of this book; particularly for their support in person. When I wrote this originally the main people who knew were my mum Rob, my brother Tim, and a handful of other people including a friend who I was lucky enough to be in a writing group with, Frank Urban. This time the support has been much wider.

My first shout-out goes to the book-club that I am lucky enough to organise/ confuse/ traumatize with my friend Emily who, after years of us both talking about there needing to be a bookclub at work, went 'Right, we're doing this' and put up a sheet of paper next to the door for anyone to write their name on if they were interested. For those who've even been slightly interested in writing- join a book club. If you've never read a book before in your life (in which case super-awesome for reading this one) and you never have any intention of reading again- join a book club. It is surprising how much fun you can have dissecting even the most morbid of books, particularly when you have a variety of strong opinions that make you see the world a little differently; before you go to one you do not realise how much you can laugh over a book, a sniff of wine, and a good serving of garlic bread. Thank you for your suggestions over the group chat, your thoughts on book covers, and that laughter in each meeting. A special throw-out to Griffin for asking some questions about the themes and focus of the book to really get me looking at it from an advertising point of view, for Anne for listening at lunch break and offering some good suggestions, Jo and Jason for your good humour, Abbey for your kindness and patience and being a good friend, and for Karlee for your statement; "nurses- we are the ultimate survivors" which prompted me to add the word Nature to the subtitle and… for the lasting image of a Continent sized seahorse- that Would be impressive. To all of you: You're all semi-marvellous (and for anyone that is uncertain, that is a Huge compliment from me).

And a thank-you for all my co-workers, for your general encouragement and support, and for Jency and Dennim in particular for having one last look at the cover- in the last few hours before submitting (who said last minute cramming finished after high school).

And a special thankyou to all my patients for being you.

Three really good friends have also been there for the journey this time, some of whom I wasn't lucky enough to know in 2016. Adam, as I have mentioned before, encouraged regularly and was one of the first to know when I said I was going to finally publish it widely; thank you also for when we went 4wdriving (when someone's driving wasn't possibly trying to impale me with a branch- I wonder who that was ☺), going mountain climbing (no fish up there but the scenery was great), and teaching me to stand-up paddle board. Carl, with your dry humour, keen focus on any mathematical problems, and reminders to keep an eye out for dolphins on the water (in case a particularly cheeky one decides to play with my inflatable- and it does not remain inflated for long) you have also been a big help- probably more than you realise. Brett, thank you for your time, your patience, and even reminding me one day to have a self-care day and to leave the book and A few other things that needed doing for a bit [multi-tasking- joy ☹], and that helped, for offering advise with the publishing (including sourcing more information, seemingly at ease, whenever more Variety turned up), being one of the first people to read the book, and even assisting with the cover. Then, on top of that, in addition to heaps of support in sourcing images, helped with transfer of this book into an epublishing file so that the rest of the world could see it- wow. You're all brilliant (and yes brilliant is on par with semi-marvellous).

Another extra special thank you for anyone else who has been a part of this in whichever way, some of you possibly without even realising it; whether that be friends, extended family, people I sing with and dragonboat with, anyone with any interests in wildlife, and the general community of Port Macquarie Hastings (including Laurieton And of course Wauchope) and my childhood home of Western Sydney: you're all quasi-awesome. Massive thanks to Claire and everyone at BookFace Port Macquarie- the first bookshop to take the chance on stocking this book which for an author is so, so cool- you're all quasi-extraordinary.

Finally, a thank you to my family. A lot of this was originally written on visits with Mum to family near Brisbane, indeed I have a distinct memory of being in my Aunt and Uncle's going through some of the various fish books that I had borrowed at the local library and gathering as many relevant facts as I could before we returned to Port Macquarie. My brother, Tim, has been there right from the beginning; listening and supporting and encouraging, in addition to being a great brother- thank you ☺. Mum in particular has been a big part of this book; being there from when it was literally nothing more than nineteen lines and a heading (not even bulletpoints), to when we posted the usb, and then later encouraging me to print it out 'after having put so much work into it' and being there to support me with that, and even framing the letter that I was lucky enough to receive. Mum also travelled with me to several museums, and aquariums, patiently waiting in Melbourne

Aquarium when I went back to have another look at the ploughnose chimaera, or indeed many other animals, whether that be venturing out onto the Great Barrier Reef near Port Douglas or looking at seahorses together in Tasmania, or even just spotting fish together as we go walking along the coast or beside the river in Laurieton. Thank you so much.

And one last especially important thank you to you for taking the time and energy to reading this; readers make books and thank you for contributing to make this one. You've even taken the effort to read the acknowledgements and That is extra-cool. Having nearly run out of complimentary terms (one-of-a-kind ones, not the boring ones you can knick from google) I can only say you're ultra-awesome. Thank you all ☺

One last final thank-you for this, hopefully final edition- fingers crossed. To the fish that proved some of the most inspiring motivation to continue on this project. The striking cowfish on the first page, for instance, was photographed in Seahorse World in a tiny town north of Launceston, Tasmania. This marine farm, housed in not much more than a large but well-looked after warehouse, is an example of little places doing big things. Their focus is the rapidly growing popularity of traditional medicine and the increased use of wildlife- including seahorses, beyond what any population of seahorse can naturally sustain. So here they farm the seahorses; allowing natural populations to continue living undisturbed lives in the wild, while educating tourists about the special little critters entrusted to their care, including other fish like this cheeky cowfish. It seemed a fitting image, in the critter's charismatic energy; even while living in an artificial home it was still showing the liveliness and adaptability that have made fish the ultimate survivors. For from whale sharks to seahorses, feisty swordfish to migrating salmon fish have found a way to make a living in the most varied of ways; living what can be remarkable lives despite their often vast differences from our own. It has been my absolute privilege writing this for you.

I hope you've enjoyed it.

Nicholas ☺

Image Details

Listed by Source, in order of their Appearances

Envato

Joe Belanger- Hunting-lizardfish.
Diegograndi- Bird Wrasse.
Pcowell- Shark feeding in sardine baitball.
Alex Vog- Shoal scalloped hammerheads.
Image-Source- Greater hammerhead shark.
Alberto Carrera- Great white shark.
Alex Vog- Grey Nurse shark.
Imagesourcecurated- Whale shark.
Diegograndi- Oscillated stingray.
ImageSourceCur- Blue Spotted Stingray
Imagesourcecurated- Spotted Eagle Rays.
Haveseen- Manta rays filterfeeding.
Rezkrr- Starlet.
Estproject - Sturgeon.
Diegograndi- Saddled Bichir.
Diegograndi- Spotted Gar
Statuska- Silver Arowana.
Sommai- Dragonfish.
TravelSync27- Tarpon.
Alberto Carrera- Moray Eel, Spain.
Dmitry Rukhlenko- Spotted Garden Eel.
Joe Belanger- Striate Anglerfish.
Edb3_16- Pacific Salmon spawning, Canada.
Galyna Andrushko- Spawning Pacific Salmon, Alaska.
Akophotography- Atlantic Salmon jumping waterfall
Beaezor- Eeltailed catfish.
CreativeNature_nl- Spotted Weatherloach.
Diegograndi- Tambaqui.
Cookelma- Red-Bellied-Piranha shoaling.
AltrendoImages- Red-Bellied Piranha close-up.
Byrdyak- Blackblotch FoxFace feeding within the coral.
Joe Belanger- Sculpin with freshly caught prey.

Joe Belanger- Hiding Stargazer fish.
TravelSync27- Dwarf hawkfish
ImageSourceCur- Bump-headed parrotfish.
Lifeonwhite- Black spotted Porcupinefish.
TravelSync27- Lionfish
Imagesourcecurated- Cleaner wrasse attending to a potato bass
TravelSync27- Mandarin fish.
Joe Belanger- Garibaldi patrolling kelp.
Joe Belanger- Sebae Clownfish cleaning eggs.
Joe Belanger- Seamoth.
GreensandBlues- Ornate Ghost Pipefish.
AddictiveStock- Seahorse.
Xcldhx038a82fd- Shrimp goby.
Xcldhx038a82fd- Sole Fish.
Diegograndi- Female Black Dragonfish.
Tan4ikk- Juvenile Trevally.
Image-Source- Oceanic sunfish.
Image-Source- Remoras attached to nurse shark.
Aarrows- Great Barracudas.
Lifeonwhite- Common Hatchet fish.
Alberto Carrera- Pipperfish.
Lifeonwhite- Four eyed fish.
Sommai- Domestic goldfish.
Dmitry Rukhlenko- Leafy seadragon.

iNaturalist

Jeffrey H Skevington- Basking Shark.
Matt Marr- Oikopleura.
Camilla Caton- Common Pyrosome.
Vasilis Stergios @vstergios78- Doliolum salp budding juveniles.
Kaila Wheatly Kornblum- Bluestriped light-bulb sea squirt.
Luca Davenport-Thomas- New Zealand lancelets.
Chris Isaacs- European Lancelet, Jersey.
Peter Southwood- Sixgill Hagfish.
Cath Rampton- Pouched Lamprey.
Philip Watson- Close-up of Port Jackson denticles.
Sabine Atteln- Sea Lamprey attached to Chinook Salmon.
Lin Ching Feng- Frilled Shark.
Yurimaoh- Frilled Shark baby.

Harry Rosenthal- Crested Hornshark eating egg.
Philip Watson- Juvenile Port Jackson.
Sam Riley- Bull Shark.
Barna Takats- White tip Reef Sharks.
Riley Wyna- Pelagic Thresher Shark.
Philip Watson- Australian Angel Shark.
Luis Pérez Berrocal - Common Guitarfish.
Patrycja Lawryniuk- Small-tooth Sawfish.
Sara Thiebaud- Spotted Chimaera.
Mandy Gardner- Ploughnose chimaera.
Daniel Zemans- American Paddlefish.
John P. Friel- Poll's Bichir.
BelowBlueWaterDiver- Eyetail Bowfin
African-Parks, Chinko, Central African Republic & Teleos- Worm-jawed Mormyrid.
Pieter Kotze- Sneaky Bulldog.
Aleksandr Levon- Elephantnose fish.
Łukasz Olszewski- Butterflyfish
Manuel V. Miranda del Solar- Pirarucu. @manimiranda (instagram). mmiranda@magia.digital (email)
Alastair Freeman- Northern Saratoga.
Hunter Stevens- migrating American Eel elvers.
Tom Vierus- Marbled Eel.
Pat Webster @underwaterpat- morays & groupers hunting
Murray Fisher- Atlantic Menhaden.
Sara Roolf- Pacific Herring eggs covering beach.
Virgil Daggert- Plainfin midshipman father egg guarding.
Andrey Ryanskiy- Fowler's Pearlfish.
Vetea Liao- Pearlfish in sea cucumber.
Peter Hancock- Jollytail Galaxias.
Taylor Davies-Colley- Koaro/ Climbing Galaxias.
Ewout Knoester- Wels catfish.
J.Martin Crossley- Beaked Sandfish.
Brothers Mathieu and David Sannier- Electric Eel.
Xaime Beiro- Fishipedia- Whiptail Catfish.
Ad Konings- Cuckoo Catfish.
Kudaibergen Amirekul- Tibetan Stone Loach,
Riccardo Novaga- European Bitterling.
Henrick M- Bluebottle Fish, Sydney.

Michal Biniek- False Cleaner Fish.
Frank Krasovec- Mating Yellowbelly Hamlet.
Harry Rosenthal- Weedy Seadragon father.
Corey Doughty- Southern Cardinalfish mouth brooding.
Alek Tritt- Blue Tilapia cichlid building a nest.
Mark Rosentein- Onefin Flashlightfish.
Harry Rosenthal- Australian Pineapplefish.
Peter Rask Møller- Sloane's viperfish.
Grant Law- Shining Loosejaw.
Marc Martin Sola- Female Black devil angler.
Gabriel Jensen- Juvenile whipnose angler.
Benoît Segerer- Tripodfish.
Marc Martin Sola- Giant Oarfish- Santa Cruz, Spain.
David Rolla- Yellowfin tuna.
Rudi Diesel- Sailfish hunting.
Justin Hofman- Blackfin Icefish.
Pam Osborn- Western buffalo bream.
Andrew Dreelin- Atlantic Flying Fish.
Matt Marr- Mexican Needlefish.
Alec Karcz- Spotted Archerfish.
Suzu Sato- Great Blue-spotted Mudskipper.
Dr John Turnbull- Red Handfish.
Tim Sorrill- Pacific Capelin washed up on shore.
Tobias Hays- Spawning Californian Grunion.
Vincent Rufray- South American Lungfish.
Sadam Toro - West African Lungfish.
Tse Chung Yi- Queensland Lungfish
Bruce Henderson, WrecklessMarine- Coelacanth.
Andrew Green- Ziebell's Handfish.
Floyd. E. Hayes- Death Valley Pupfish courting.
Björn Källström- Stranded Basking Shark

Wikipedia

National Ocean and Atmospheric Administration (NOAA) Fisheries Service, Greg Skomal- Basking Shark.
Insert from Walcott Cambrian Geology and Paleontology II- Pikaia.
Monterey Bay National Marine Sanctuary, Linda Snook- Pacific Hagfish.
Marine Fisheries Review, Karsten Hartel - Cookiecutter shark.

NOAA Okeanos Explorer Program, Gulf of Mexico 2012 Expedition- Longnose Chimaera.
Nouvelles Archives du Muséum- Chinese Paddlefish.
Spix, J.B.V. and L. Agassiz - Agassiz's Arapaima.
Sandra J. Raredon, Smithsonian Institution, National Museum of Natural History, Division of Fishes- Fangtooth.
Brauer, A., 1906- Jelly faced Spookfish.
NOAA Ocean Exploration- Danae Viperfish.
G. Brown Goode & Tarleton H. Bean, "Oceanic Ichthyology" 1896- Elephant Eel.
U.S. National Oceanic & Atmospheric Administration- Rough Abyssal Grenadier-
Alaska Fisheries Centre, National Marine Fisheries Service, 2008 Beaufort Sea Survey, Erica Acuna- Careproctus Snailfish
Vailulu'u 2005 Expedition NOAA-OE- Cutthroat eels.
NMFS-PIRO Observer Program- Juvenile Oarfish.
Jean-Paul Cicéron, Mangrove Killifish.
Zina Deretsky- National Science Foundation (USA)- Reconstruction of Tiktaalik.

Other

Nicholas Kable- Inside Cover- Shaw's Cowfish (Seahorse World, Launceston- Tasmania). Dunkleosteus- Brisbane Museum.
Timothy Farlow and Nicholas Kable- Bristlenose Catfish- Home Aquarium of Timothy Farlow, Brisbane.
Chris Crippen - Muskellunge.
Chris Crippen- Baby Seahorse.
Woods Hole Oceanographic Institution- Pink Ventfish.
Nicholas Kable- Finis- Indo-Pacific Sergeant major- Great Barrier Barrier Reef.

Index of Fishes

Finis

About The Author

Nicholas Kable is originally from Western Sydney (Quakers Hill) and has called Port Macquarie home since childhood. He has experience in health, initially as a volunteer in aged care, since 2003, and was also recognised with a Certificate of Appreciation on International Volunteers Day (2011) due to his contribution to Alzheimer's Australia NSW. He has been published as a contributing writer in the Hastings Writers anthology 'My Wonderful Hastings', and was the co-author of the play on 'Justice Wilson Hudson' which was performed at the Port Macquarie Historic Courthouse. His main writing and researching interests are history (particularly Australian) and wildlife. 'The Life of Fish; Nature's Ultimate Survivors' is his first book as a solo author.

www.ingramcontent.com/pod-product-compliance
Lightning Source LLC
LaVergne TN
LVHW052241200926
848481LV00002B/235